ANIMALS AS NEIGHBORS

THE ANIMAL TURN

Series Editor

Linda Kalof

Series Advisory Board

Marc Bekoff, Juliet Clutton-Brock, Nigel Rothfels

Making Animal Meaning
Edited by Linda Kalof and Georgina M. Montgomery

Animals as Domesticates: A World View through History
Juliet Clutton-Brock

Animals as Neighbors: The Past and Present of Commensal Species
Terry O'Connor

ANIMALS AS NEIGHBORS

The Past and Present of Commensal Species

Terry O'Connor

Michigan State University Press
East Lansing

∞ The paper used in this publication meets the minimum requirements of ANSI/NISO Z39.48-1992 (R 1997) (Permanence of Paper).

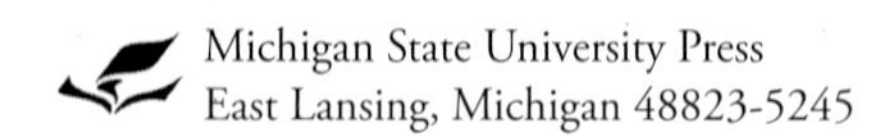
Michigan State University Press
East Lansing, Michigan 48823-5245

Printed and bound in the United States of America.

19 18 17 16 15 14 13 1 2 3 4 5 6 7 8 9 10

LIBRARY OF CONGRESS CATALOGING-IN-PUBLICATION DATA

O'Connor, Terry.
Animals as neighbors : the past and present of commensal species / Terry O'Connor.
pages cm — (The animal turn)
Includes bibliographical references and index.
ISBN 978-1-61186-095-5 (cloth : alk. paper)—ISBN 978-1-61186-098-6 (pbk)—
ISBN 978-1-60917-387-6 (ebook) 1. Human-animal relationships—History. 2. Commensalism. I. Title.

QL85.O29 2013
577.8'52—dc23 2012049443

Book design by Scribe Inc. (www.scribenet.com)
Cover design by Erin Kirk New

green press INITIATIVE Michigan State University Press is a member of the Green Press Initiative and is committed to developing and encouraging ecologically responsible publishing practices. For more information about the Green Press Initiative and the use of recycled paper in book publishing, please visit www.greenpressinitiative.org.

Visit Michigan State University Press at www.msupress.org

To my family (and other animals)

Contents

Preface

Among my strongest childhood memories are times spent on a seaside promenade, throwing pieces of stale bread into the air to be caught by swooping, squabbling gulls. Birds that could be seen at other times soaring over the waves or poking about in rock pools would come to within a few feet to aerobatically snatch scraps of food, temporarily leaving their "wild" world to engage with ours. To a child, there is something quite special about feeding a wild animal. As adults we might analyze and rationalize the process in terms of conservation, of mitigating the impact that our homes and offices otherwise have on "nature," but something of that magic remains. Whether we feed them deliberately, or they feed themselves opportunistically from our leavings, or find shelter in the structures that we thought we had built for ourselves, the everyday animals that live around our homes and workplaces interdigitate their lives with ours and occupy a particular place in our attention. That alone would make them deserving of study, but we humans have been planting our footprint across the planet for millennia. How have other animals responded to that impact: as a threat or an opportunity? In short, how long has this been going on?

This book presents research that meanders along the boundary between biology and archaeology, sometimes focusing on what we know of the ecology and behavior of our present-day neighbors, and sometimes delving into the historical and archaeological records. Either approach would be incomplete without the other. To understand the contemporary relations between ourselves and our neighbors, we need to understand the time depth, and the consistency through time. And to understand the animals whose remains we find around archaeological sites, we need to understand those animals as living creatures with their own knowledge, skills, and agenda. Researching the book has been both a challenge and a joy. The literature is vast and dispersed across academic works, journals, blogs, and assorted mass-media formats. The reach of the book is global, as the subject cannot be constrained by geography, and I am grateful to have had the excuse to read my way into some unfamiliar regions and environments. Above all, though, the book is about animals. I have started from the premise that other species are intelligent and resourceful, and that in the prehistoric past, our ancestors were our equals in intelligence and in the curiosity and attachment that they may have shown to the animals that were attracted to our earliest settlements. Urban foxes may be nothing new; even feeding the birds may be nothing new. Noticing those animals and drawing them into our social and cultural realms is so widespread among people today that it, surely, cannot be a modern phenomenon, hence this exercise in integrating biology and archaeology. There are some thoughts for the future, too, as it is the adaptable animal neighbors who are coping most successfully as we build and farm over ever larger parts of the Earth. What is it that enables some species to succeed, and what can we do that might enable more species to do so? The latter question is not simply altruistic. We need animals in our lives, sometimes iconically and symbolically, sometimes for their therapeutic role, and our neighboring species are the ones that occupy those roles in most people's lives, most of the time.

I am very grateful to Linda Kalof for encouraging me to write this book, and for her good-natured patience, and to Annette Tanner, Julie Loehr, and their colleagues at MSU Press for seeing it into print. My thanks also go to those who read and commented on earlier drafts—in particular an anonymous reviewer for MSU Press, and James O'Connor—for asking the right questions at the right time about the early chapters. Thanks go to all who have allowed the use of images, individually acknowledged in figure captions, and to Museum of London Archaeology for data on rodents in chapter 5. Above all, I thank my family for putting up with me, and the many commensal animals that have kept me company during the writing of this book and who, quite literally, work for peanuts.

Introduction

It hardly needs saying that animals are central to our individual and collective lives. As utilized livestock, they feed us and provide other resources such as wool and leather. In arid lands, they convert the poor vegetation of rangelands into animal protein, enabling some living to be made in otherwise unproductive places. In less-industrialized economies animals provide power, hauling carts and plows, again making agriculture possible. In our homes, animal companions become a part of the family, an outlet from the complexity of human relations, and a way of teaching our children about responsibility and care. Domestic livestock have been with us for around 10,000 years, replacing the uncertainties of hunting by bringing animals under direct control.[1] Pets and companion animals are more difficult to trace in the past, though the association of people and dogs predates the domestication of livestock.[2]

So important and familiar have livestock and companion animals become that it is easy to assume that all such relations have arisen because people willed it. We see the emergence of animal husbandry and its subsequent elaboration as "stages" of human cultural development,[3] instances of people taking ever more control of their lives and of the world around them, in parallel with the emergence of more complex metalworking and other pyrotechnologies. As our cultures elaborate and diversify, so animals come to be seen, at best, as one of the media through which human economic and power relations were mediated, as part of the props and scenery of the human drama, not as actors in their own right. It is all too easy to forget that those other species are adaptable, resourceful animals with drives and agendas of their own. For some of them, the global spread of one noisy, untidy, demanding primate has been little short of an unmitigated disaster. The archaeological and paleozoological records are rich in examples of species that have been driven to extinction, locally or globally, either by the direct actions of people or by the indirect consequences of our habits of constructing and modifying the world around us.[4] Other species have successfully adapted to us and our ways, seizing the opportunity that our planetary dominance presents, greatly increasing their numbers, and extending their range beyond what was possible in the absence of people.

This book is about the ways in which many species of animals have contrived to gain some benefit from living alongside people within the constructed and heavily modified environments that we make for ourselves, and the long history and prehistory of that process of cohabitation. In order to explore that relationship, it will be necessary to discuss human-animal affiliations more generally, and to question the terminology that we apply to them. It will be clear by now that "animals" refers predominantly to vertebrate animals, and to mammals and birds especially. That is not to say that invertebrate animals and microbes are somehow less significant: far from it. Search the scientific literature for the term *commensal,* and the great majority of results will be papers dealing with the microbiota that populate our bodies. However, if we want to understand the past status of our neighbor species, we need to study those that leave substantial traces in the historical and archaeological records, and if we are interested in the

Figure 1. A hard-working donkey in Luxor, Egypt, and a much-loved family pet, both are indisputably domestic animals, though they have very different relationships with the people around them. (Source: author.)

diversity of cultural responses to other species, we need to limit our study to those groups—mammals, birds, the larger herpetiles—that are conspicuous enough to have prompted some response. We are all familiar with such species: we see them every day. Exactly which species we see will depend upon where on Earth we live, though some will be familiar everywhere. My own animal neighbors include some that would be familiar to a visitor from overseas (house sparrows *Passer domesticus*, cats), some from animal families that have commensal members worldwide (mice), and some that have locally adapted to a commensal life (bullfinches

Pyrrhula pyrrhula). What is consistent, right across the planet, is that close association of our species with others. They have adapted to our living space, either because we have extended that living space across more and more land area, forcing species to adapt, depart, or die out, or because we have provided feeding and living opportunities that have encouraged some species to move into our territory. A few species have become our neighbors somewhat grudgingly, managing to persist in our living space, yet sensitive to further constraint or interference, while others have moved in with apparent enthusiasm and with conspicuous success. Some of our animal neighbors have lived alongside us for centuries or millennia, while others have adapted to our living space comparatively recently.

Figure 2. The dodo *Raphus cucullatus* was a species famously maladapted to association with people and is now known from a few bones, stuffed bodies, and illustrations. (Source: Michael Leveille, Biodiversitymatters.org, with permission.)

One of the aims of this book is to look back over the past of humanity and explore the time depth of this mutual adaptation; although the text deals with commensal animals, it is

Figure 3. Konik horses and Heck cattle are probably the closest surviving representatives of wild horses and cattle, species that have been highly successful as domestic livestock while becoming extinct as truly wild populations. (Source: Hans Kampf, Large Herbivore Network, Oostvaardersplassen, Lelystad, the Netherlands, with permission.)

Figure 4. A house sparrow *Passer domesticus* looking for crumbs, and urban herring gulls *Larus argentatus* looking for trouble are uninvited neighbors. (Source: author.)

about commensalism. As such, the book is about people as much as it is about animals: as we shall see, the two have been inseparable for millennia. We not only provide the commensal opportunities but also decide which of the opportunist neighbors can be tolerated, which accommodated, and which reviled. A full treatment of the anthropology of "vermin" would fill a separate volume, though it is a subject upon which we touch a number of times. The point is that we cannot understand the emergence of these commensal adaptations without understanding the human context, nor can we usefully discuss the place of those animals in people's lives now or in the past without understanding the ecology and ethology of the species concerned. Thus, a number of chapters explore the present-day biology and behavior of our animal neighbors, while others are essentially archaeological, exploring the origins of our close cohabitation with other species, how that has changed over the long course of time beyond the reach of written history, and thus how we have reached the current situation. Just as the archaeological study of a geographical region will include a detailed survey of the landscape as it is today, so this study of the archaeology of our animal neighbors must include detailed examination of the present-day ethnozoological landscape. And conversely, in order to understand the present-day landscape, metaphorical or actual, it is important to consider how it came about.

This concern with integrating ecology, field biology, and archaeology reflects the author's research interests and personal disposition. I have spent over thirty years working in the field of *zooarchaeology*: studying the associations between past people and other animals through the surviving material record. Zooarchaeology arose in the mid-twentieth century as archaeologists realized how much important economic and environmental information could be extracted from the bone fragments recovered in quantity from archaeological deposits of all periods, worldwide. In its formative years, zooarchaeology tended to be the province of zoologists and paleontologists who applied their expertise in the identification of old bones. Gradually, however, it has come to be the research discipline of archaeologists who specialize in animal bones, much as others might specialize in Roman pottery or Mayan glyphs. Where the main questions were formerly "Which species, where, when, and in what numbers?," we are now more prepared to ask, "What did these animals mean to those people?" Nonetheless, there continues to be a certain narrowness of outlook in zooarchaeology. Decision making and cultural adaptation are the prerogative of the ancient people concerned; the animals passively ate, mated, perhaps migrated, and died. People are the subjects of zooarchaeology, animals too often the objects. Yet it is only necessary to spend time in a leafy urban park to see the behavioral adaptations of squirrels, pigeons, crows, sparrows, and dogs to the people around them and to each other. Furthermore, we might watch that same range of animals in a park in New York, London, or Warsaw, which should give further food for thought. My starting point in writing this book has therefore been to subvert one of zooarchaeology's research questions by asking, "What did those people mean to these animals?"

One of the challenges that we face is that of appropriate terminology. By "appropriate," I mean sufficiently precise and unambiguous as to minimize questions of intended meaning. In the Anglophone world, as in many other widely used languages, we make a simple distinction between wild animals and domestic animals, a dichotomy that does not reflect the range and subtlety of human-animal associations. This is problematic because we use such terminology to categorize the world around us in general, and animals in particular. That in turn affects how we behave towards that biota. A key piece of conservation legislation in the United Kingdom is the 1981 Wildlife and Countryside Act. But where exactly does "wildlife" begin and end? House mice and street pigeons are certainly not beneficiaries of the Act in the same way as dormice and golden eagles. In fact, the Act has nine distinct schedules, each of which has numerous parts, giving different forms and degrees of protection to different categories of "wildlife."

Figure 5. Cattle and sheep bone fragments among occupation debris from an archaeological site in York, UK. Animal remains form an abundant and informative part of the debris of former human lives. (Source: author.)

A similar problem arises when we consider human impacts on global biodiversity. This impact is generally seen as ranging from negative to catastrophic, yet many species have greatly increased their global populations and distribution over the last few centuries. Some of those species are the domestic livestock whose numbers and breeding are largely under direct human control, but many other species have also spread and multiplied by adapting to the habitats that we build for ourselves. We have an ambiguous relationship with these species, preferring not to include them when we discuss "wild" animals, yet accepting that they cannot be classed as "domestic." The importance of our familiar neighbors in terms of biodiversity and of environmental education is discussed in a later chapter. The same problematic issues arise when we look back into the archaeological past. Here is species X, its remains identified at a certain time and place; how are we to explain its presence and ecological role? Our default starting, and often finishing, position is usually to consider how we would categorize that species today, and to take that as the answer to our question unless there is overwhelming evidence to the contrary. A nice example is the white-tailed sea eagle *Haliaeetus albicilla* in the archaeological record in England. Today, this magnificent bird is redolent of Scotland's wildest coastal places, a reintroduction and conservation success story. In previous centuries, at least into early medieval times, it seems to have been a quite widespread species, one of the birds that scavenged around human settlements. From village scavenger to majestic symbol of "wildness" in just a millennium.

Figure 6. White-tailed eagle *Haliaeetus albicilla* over an Anglo-Saxon village. Though a bird of wild rocky coasts today, these eagles seem to have been common scavengers around Saxon and medieval settlements. (Source: Illustration by Julie Curl, Sylvanus Archaeological Natural History and Illustration Services, with permission.)

We are concerned with describing and understanding the associations that arise between ourselves and other species. Elsewhere in the biosphere, close association of two or more species is quite common.[5] Although we might simplistically believe that other animals interact by mating, by eating each other, or by ignoring anything that is neither mate, predator, nor prey, field biology shows that species interact in far more complex and subtle ways. The terminology outlined here is that which has been found to be necessary and useful for this book; it is not intended to be definitive or exclusive.

Biologists use the term *commensal* to refer to species that adapt to the living space of another, and call the adaptation *commensalism*.[6] The term derives from "together at table," reflecting the fact that commensalism generally conveys some feeding benefit. The most obvious commensal organisms, already mentioned, are the myriad microorganisms that inhabit the human alimentary canal. Some are symbionts, conveying a degree of positive benefit to their human carrier, while others simply find the gut to be a congenial home from which they benefit, but in which they do neither good nor harm. These species should not be confused with parasites or parasitoids, which spend all or part of their life span feeding directly off a host species, often within the host's body, and frequently to the detriment of the host. Intestinal worms are parasites of cats, and cat fleas are parasitoids of cats and (too often) of people; however, cats are commensal with people, not parasites, however much we may sometimes think so. Such is the diversity of animal behavior that behavioral terms tend to overlap at the margins, so there may be some question, for example, over the status of bird species that hang around the breeding colonies of other birds, opportunistically feeding off food scraps left by the host (commensalism), off food that the host species has been harassed into regurgitating (kleptoparasitism—parasitism by theft), and off nestlings that are inadequately protected (predation).[7] Nonetheless, we have to define our subject somehow, so for the purposes of this book, *commensal* is taken to refer to those animal species that utilize the modified or constructed environment of human habitations for living space and food. Some of those species may spend a part of the year or of their individual lives away from human habitation, living "wild" in a conventionally understood sense. The key attribute is that habitats constructed or heavily modified by people either provide the majority of living space and subsistence, or provide those resources at a critical point in the animals' lives, without which their populations would not be viable. Commensalism is, for the purposes of this book, an emergent property of the human trait of modifying and constructing living space, further discussed in chapter 1. Many species do this to some degree—consider a termite mound—but no other species modifies its living space to such a degree, in so many different ways, nor throughout such a range of biomes.

Another term that is widely applied to such species is *synanthropic*—literally "with people."[8] Synanthropes are organisms that live in and around human settlements, so the term may appear to be a synonym of commensal as used here. However, the latter term, with its meaning implying sharing of food, more nearly reflects the importance of anthropic environments for food as well as for living space. An important insight in the emergence of ecology as a mature science was Hutchinson's model of ecosystems as being structured by energy flows and stores. In an ecological study, then, the trophic relations of people and other animals are central to sustaining their relationship, and as will become clear, it is access to food that drives the great majority of the commensal relationships discussed here. If we study the relationship between biodiversity and ecosystem stability in human-dominated environments, we find that the details of food webs, the trophic relations of who eats what (or whom), are fundamental to understanding how such environments function.[9] For those reasons, this text uses *commensal* and its derivatives in preference to *synanthrope, synanthropic*, while accepting their near synonymy. Commensal living could be seen as a specialized form of being synanthropic: not only tolerating the close proximity of people, but also gaining benefit from the access to food that results from the proximity.

There are obviously shades within this category. Some species live largely or entirely within our buildings (cockroaches, house mice), while others live largely in the modified environment around and between our structures (house sparrows). Ecologists have proposed a range of terms to make distinctions among commensal or synanthropic species. Michael McKinney[10] summarizes the responses of other species to our living space as threefold: exploiters, adaptors, and avoiders. These three groups coincide neatly with Johnston's categorization of full synanthropes, casual synanthropes, and non-synanthropes.[11] Within McKinney's exploiters, a further subtlety separates species that live within the buildings in which people live, sometimes termed *domiciliary* species.[12] All of these terms have porous boundaries: a commensal rodent that is domiciliary in winter may move out of the house but remain within the immediate built environment during the warmer months. How do we regard such species? Are they household vermin that seasonally leave us alone, or are they wild animals that seasonally move in with us? We are dealing with a wealth of different strategies for food and living space that grade into "wildness" at one extreme and familiar domesticity at the other. What connects those strategies is that the species concerned depend upon, or utilize, or exploit (you decide) people and our environmental modifications as a resource base for their own ecology. The very fact that we need to define and accommodate such a nuanced diversity of lifeways is a cogent reminder of the adaptability and versatility of other species. The sheer diversity of the terminology involved is also another nail in the coffin of a simple wild/domestic dichotomy.[13]

There is a whole other category of animals that lives closely with people, of course—namely, the ones that we call "domestic." This book is not about them, but a brief digression is helpful in order to establish boundaries, albeit shifting and porous, and to discuss the ways in which some familiar terms are used. A great deal of ink has been shed in the quest to define "domestication."[14] Some have attempted a largely zoological definition, characterizing domestication as a form of speciation event, by which a "wild" species became a "domestic" species. This view is reflected in the formal decision to accept and apply binomial ("Latin") species names to domestic animals: thus the domestic ox *Bos taurus* is given a different taxonomic name to its ancestor species *Bos primigenius.*[15] In essence, this approach sees domestication as an event. At some point in the past, people deliberately took populations of animals into close management and so effected the division of the new "domestic" species from its wild progenitor. The purpose was to gain the benefits of close control of the animals, but the consequence was to initiate speciation through isolation and changing selection pressures. Others have taken a more sociological view and emphasize the incorporation of the animal into human social structures; "domestication" thus becomes a process of taking the animal into the *domus,* though it remains an outcome of human decision making. Still others, including the present author, have characterized domestication as just one part of a spectrum of affiliative relations arising from the coevolution of people and other species—relations that are subject to ongoing negotiation and change, and not all of which are necessarily intentional.[16] Although the present volume defines its subject matter by excluding domestication, even that is problematic if "domestication" itself is not well defined. We would have little difficulty in agreeing that farm livestock are domestic animals, and they are not discussed in this book. However, consider the humble pigeon *Columba livia*: it may be a wild animal (as free-living rock dove), a commensal animal (as an urban scavenger), or a domestic animal (as homing or "fancy" pigeons). Although we tend to apply these labels at the species level (sheep are domestic animals, lions are wild animals), they should more properly be used in reference to populations within those species. The importance of categorizing populations, and not generalizing across species, is a theme that recurs throughout this book and is discussed at length in chapter 7.

Two other terms require definition and brief discussion. *Feral* is one of those handy words that has both a biological and a vernacular meaning. For this volume, "feral" will be taken quite strictly to refer to populations or individuals originating in a formerly domesticated population, but that now live independently of people—i.e., that have "gone wild." The more relaxed use of "feral" as a synonym for "wild" is not helpful: journalists and politicians occasionally refer to "feral children," a derogatory term that often wrongly implies they were domesticated in the first place. The narrower definition of "feral" is useful, and will be used here. Although *tameness* is sometimes regarded as one defining characteristic of domestic animals, a *tame* animal is not necessarily a domestic one,[17] and this term should be applied at the individual level. Many animals can be tamed to some extent if habituated to people at an early age. Individual big cats can be tamed, but by no stretch of the vernacular taxonomy could leopards be described as "domesticated." A specific leopard may, however, be a pet or companion animal, so these two terms cannot be equated with "domestic" or used in opposition to "wild." Furthermore, when we try to understand human-animal relations in the past, we cannot be confident that the categories we define today, however inadequately, would have been applicable in earlier times. Thus we may fall into serious category errors if we make distinctions between, for example, wild and domestic *species* rather than *populations*, or assume that the relationship that is currently typical of people and another species is necessarily the long-term default condition, rather than a currently adaptive but temporary relationship.

Figure 7. A tame large felid, perhaps a cheetah, illustrated on the New Kingdom mortuary temple of Hatshepsut at Deir el-Bahri, Upper Egypt. (Source: author.)

There are terminological pitfalls to be avoided, then. Of the many terms that are used to categorize other animals or to describe their relationship with people, some can only be applied to individual animals, others to populations. Whether any can satisfactorily be used at the species level is debatable. We might briefly agree with the assertion that "Killer whales

Orca orcinus are a wild species"—until, that is, we see local populations of them following fishing boats and feeding off the disturbed fish and bycatch. Does that make them commensal? Not for the purposes of this book, as those opportunistic killer whales are not moving into human living space, but are merely taking advantage of opportunities afforded them by our own feeding behavior when we temporarily move into their living space. This is *facilitation*, yet another interspecies relationship, and one that is remarkably widespread among other species. A nice example is the way that some cormorant populations have learned to follow rays (Rajidae) when the fish are feeding in shallow water, scooping up the marine life disturbed by the ray as it scoots along the bottom.[18] Any discussion of commensal animals must steer a course between facilitation and domestication, and this volume will certainly overlap a little in both directions. However, the aim is to focus on those animals that share our space and our food of their own volition. These are the volunteers, the adaptable opportunists that did not wait to be invited to share the world that we built for ourselves.

The capacity of other species to adapt their behavior so as to take advantage of specifically modern, urban habitats has been noted by some ecologists and given the name *synurbization*.[19] Successful synurbic populations show a particular range of adaptations, including increased population density, reduced territories and migratory patterns, increased longevity (albeit often with reduced health status), and diversified diet. As Maciej Luniak points out, much of this is within the behavioral plasticity of the species concerned; it is the attainment of a highly specialized, realized niche through constraint of the broad fundamental niche of the species. Synurbization is certainly a relevant concept for our present purposes. However, it is generally described as something of recent origin, an outcome of other species adapting, of necessity, to the proliferation of modern cities. The aim of this book is to consider the past as well as the present, and so to consider the adaptation of other species to human settlements that had some of the attributes of modern cities (e.g., garbage, buildings) but not others (e.g., road traffic, street lighting). The pigeons of present-day New York can be described as synurbic, but it would stretch this useful term beyond breaking point to apply it to the house mice of Neolithic Çatal Hüyük (chapter 5).

Figure 8. Urban peregrine falcon *Falco peregrinus* photographed in central Edinburgh, UK. (Source: Peter Cairns, Northshots.com, with permission.)

What makes a successful commensal animal? The answer will depend to some extent on the characteristics of different human settlements and contingent historical factors as much as it does on generalizations regarding the physiology and ecology of individual species. Omnivory, or at least a broad-spectrum diet, is clearly advantageous. A species that is going to be commensal with people needs to be able to extract food value from the wastes and refuse that people deposit. Obviously that will be highly variable through time and across space, but is likely always to have included a mix of plant- and animal-based foods. More to the point, stored foods, spillages, and refuse will attract and develop a diverse fauna of invertebrate animals.[20] A species that has the ability to prey on those invertebrates as well as on the debris of human activities will obviously be well adapted. Ultimately, human settlements develop their own food web, driven in part by the primary productivity of whatever green plants manage to colonize gardens, waste ground, and midden heaps, but mostly by the energy input represented by food waste and other refuse.[21] Human settlements become energy "sinks," drawing in energy, embodied as food and fuel, from a large catchment area, then expending that energy within the smaller, busier confines of the settlement area, or storing it in concentrations of people and other living organisms, or as energy-rich refuse. There is a niche for some top predator in this food web, and we will see examples later in this book. However, it is the omnivores that are likely to be most successful, feeding at several trophic levels and able to adapt to whatever handouts human activity provides.

Adaptability, then, is important, but not only in terms of diet. In order to exploit the feeding opportunities, it may be necessary to overcome some behavioral characteristic that is otherwise typical of the species but that would be maladaptive. A good example is the tendency towards intraspecies antagonism in cats: wild cats (and many house cats) aggressively defend territories, yet feral cats coexist in surprisingly close proximity to one another, adopting social structures that reduce antagonism.[22] By doing so, they can better exploit concentrated and localized feeding opportunities. By being more tolerant of each other's proximity, cats benefit individually, gaining access to food sources that they might otherwise have avoided. Flexibility of breeding season may also be a beneficial adaptation, enabling a commensal species to align its breeding biology with seasonal changes in resources and opportunities.[23] Conversely, the year-round availability of living space and food may allow a species to broaden its breeding repertoire, maintaining breeding conditions for a greater part of the year and with less need to time births to benefit from seasonal food sources. On the same point, a successful commensal species will probably have a high rate of reproduction, enabling it to build up and, if necessary, replace populations rapidly when circumstances permit. All of these generalizations are, of course, subject to local circumstances of time and space, and should not be taken to indicate that no slow-breeding carnivore could ever successfully establish commensal populations. The increasing presence of peregrine falcons *Falco peregrinus* in our cities is a timely reminder of the danger of such generalizations.[24]

So much for biology; what of the human aspects? Commensal species are the animals that feature in people's daily lives, whether as agreeable companions, undesirable vermin, or merely part of the everyday scenery. That fact alone gives them a particular significance. They are not commercial animals, utilized by people as subsistence and other resources, traded as commodities, raised and slaughtered as necessary. Nor are they truly wild or free-living, conducting their lives far distant from ours and coping only poorly when their living space and ours intersects. Different cultures obviously categorize animals in quite different ways, and the distinction made here between wild and domestic, or between commensal and free-living, is a distinctly Western, twenty-first-century categorization, at least arguably derived from a Judeo-Christian ethic with a respectful nod to St. Thomas Aquinas. More mundanely, we classify animals according to the places in which we encounter them. The taxon "garden birds" is frequently encountered in the UK, even though it has no biological or legislative basis.[25] Hunter-gatherer peoples of recent times have held a range of attitudes towards the "wild" animals on which they depended, often with no concepts of ownership or tameness, or even a clear distinction between people and other animals.[26] The identification of the hunter with the prey is a frequent theme in folk stories, and many cultures invoke or represent beings that are an amalgam of human and animal features. In Paleolithic Europe, for example, some of the earliest carved objects, other than tools of some form, are of what appear to be human figures with the head of a lion.[27] Such theriomorphic imagery reached a peak in Ancient Egypt, with numerous deities depicted as essentially human postcranially, topped off with the head of a lion, an ibis, a jackal, and so on. The transference of attributes between people and other species (sometimes called *therianthropy*) is a common theme of many human societies, and it pulls the rug out from under our classifications of other animals. Part of the interest of commensal animals lies in our attitude towards them, how culture-specific those attitudes may be, and how they may have changed in the recorded past. We coexist with these neighbor species today, and it is at least a reasonable presumption that we did so deep into our, and their, past. Commensal animals must have featured in human lives, to have been a part of our ecology and our everyday landscape, for thousands of years. Hence the two linked perspectives of this book: to look around at commensal animals today, and to look back at evidence for our cultural coevolution in the past.

Perhaps, too, we may look forward. One of the key things about commensal animals is that they are successful. In a world in which the human impact on other species is so often seen as destructive, it is good to be reminded of the remarkable number and diversity of species that have found a successful way of living close alongside us. There is deep cause for concern in the loss of biodiversity around the world, and pigeons may be a poor swap for Amur tigers, but we should not allow acute conservation concerns to blind us to the success with which some animals have adapted to us and to our constructed environments. The world needs its wild places—in part as samples of major biomes and reserves of biodiversity, and in part because psychologically we all need to know that those wild places exist. However, we must not privilege "the wild" to such an extent that we fail to explore and to learn from the familiar. Within wildlife conservation, there is an increasing realization that towns and cities may be managed as "patches" of distinctive biodiversity.[28] Architects and planners explore new ways to make our living space more amenable to other species, and there is a growing realization that the educational and therapeutic value of contact with other animals can be most effectively realized in urban settings.[29] In all of these ways, it is the familiar commensal animals, the species that are under our noses day after day, that are the success stories that may help us to find ways to mitigate our impact on other species and other landscapes. And by exploring the long story of that mutual adaptation, we can come to appreciate those species for being as much a part of our cultural heritage as any artwork, temple, or burial mound.

1

The Human Environment

To speak of the "human environment" may seem redundant. There are few terrestrial environments where some human influence or modification cannot be discerned. We are making small but significant changes to the composition of the atmosphere, major changes to the ecology of all but the deepest oceans, and we have radically altered the composition and distribution of plant and animal communities all across the Earth's land surfaces. Human activities are now a major factor in soil and sediment erosion in many parts of the world. In the more densely occupied regions, sediments directly resulting from human activity are a significant facies of the superficial geology, attracting the attention of us archaeologists.

So widespread and distinctive is this impact that some scientists have proposed the term "Anthropocene" to define the current period of the Earth's history, in which the human "fingerprint" can be detected in most geological and atmospheric processes.[1] The Anthropocene is certainly useful as a concept (and a better term than some of the alternatives that come to mind—the Tarmaciferous, the Machinian?) and is becoming widely used before its formal adoption as a geostratigraphic unit. There is little question that we are *in* the Anthropocene, but when did it begin? Different authorities have proposed the eighteenth century A.D., linking it with the Industrial period,[2] while others have pointed out that human environmental impact began much earlier, perhaps as farming spread around the world.[3] Either may be an appropriate definition, depending on the scale at which we examine those environmental changes. From the perspective of human-animal associations, what mattered was the point at which human impact on the local environment disrupted existing ecologies and created new opportunities. The timing of that point will have been locally contingent, a key element being the construction of permanent human settlements.

In the modern world, the majority of people live in towns and cities: huge, complex artifacts that retain little of the original landscape other than as modified fragments that constitute isolates and corridors within the human construction. To humans, towns are a tool by which we shape our surroundings and facilitate our social and economic activities.[4] To other species, our towns and other settlements are both a challenge to their adaptability and an opportunity. The nature of those challenges and opportunities will be, to some degree, locally contingent: every settlement will have its own particular attributes. However, we humans have enough needs and cross-cultural behaviors in common that we can, perhaps, generalize usefully about the challenges that will face most commensal species.

From the outset, we should be careful not to regard environmental modification as a uniquely human trait. Most species alter the environment around themselves to some extent.[5] As parallels to our own towns and cities, we are familiar with the complex nest systems and

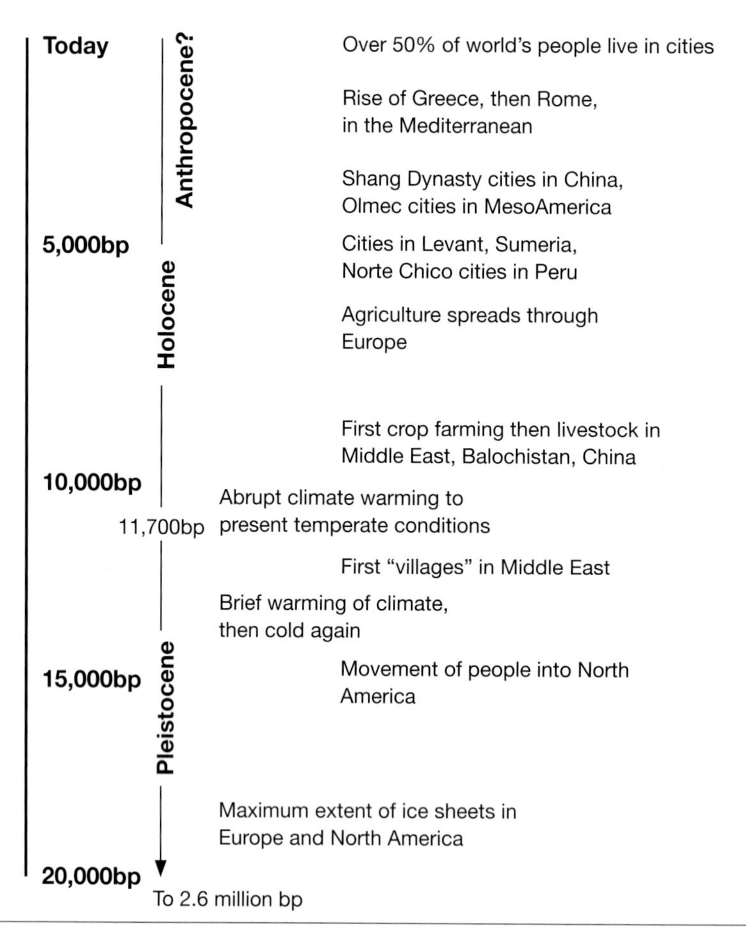

Figure 9. This timeline shows the emergence of villages and towns. The division between Pleistocene and Holocene is conventional. The status of the Anthropocene is a matter of current debate.

fabricated "mounds" of ants and termites, which not only consist of the nest construction but also often show a surrounding environment heavily modified by the feeding and cropping activities of the insects. In marine ecosystems, colonies of invertebrate hydroids secrete for themselves elaborate calcareous structures that cover many square kilometers with coral reefs, and which will provide a living for numerous commensal species. More spectacular still is the environmental change wrought by green plants, which have permanently altered the Earth's atmosphere by dumping into it their metabolic waste product, that highly reactive and dangerous gas oxygen. We humans cannot match that achievement, despite our best (or worst) efforts, though we have constructed numerous patches of wholly fabricated habitat—our houses, towns, roads, and so on—and substantially altered the environment around those constructions. It is easy to regard a road or a new town as environmentally destructive, but what such constructions do is to replace one environment, one set of habitat patches, with another.[6] That new environment may not be "natural" in any sense, but that does not mean that it is not attractive to, and a viable home for, quite a range of species. We may not be the only species that actively modifies its environment, but we are certainly the only one that worries whether or not that environment is "natural."

A pedantic digression is required here to consider what we mean by "environment." Like so many of the terms that have found their way into the discourse of contemporary politics, environment has acquired a number of subtly differentiated meanings. Perhaps the simplest, and the most useful for us here, is to say that the environment of a population of people or other animals is the physical setting in which that population exists, defined in space by the range of the population concerned, and subsuming a diversity of descriptive parameters. In that sense, "environment" always has to be qualified: the environment of who or what? The journalistic tendency to refer somewhat hand-wavingly to "the environment" is not helpful here. So as I write this chapter, my environment consists of a breeze-block constructed garage-cum-office that is situated within a small patch of highly diverse herbaceous and shrubby vegetation that is mostly alien to the geographical region. Beyond that small patch is an extensive landscape of undulating grassland maintained by grazing farm livestock, interspersed with clumps and lines of trees and several small streams. The climate is cool-temperate, with year-round precipitation averaging around 800mm per annum and prevailing westerly winds. I could go on, adding detail such as the cats and numerous frogs that share the same patch, the seasonal influx of butterflies, the significance of neighbors' offspring as an environmental noise factor. The point is that "my environment" consists of several different, nested, but somewhat overlapping zones. At the core is a built environment, a structure within which subsistence and other activities take place, which consists of a mix of manufactured (breeze-block, concrete) and "natural" (wood) materials, but which would not exist without human deliberation and manufacture. Within that built environment there are patches of different habitats (dusty crannies where the woodlice—Isopoda or pill-bugs—live, well-lit window corners favored by spiders). The building is a distinctive environment, but quite heterogeneous at a small scale. Beyond and around the buildings is space in which daily activities take place, and which has been heavily modified to meet a priori needs. The garden could be regarded as manufactured, part of the built environment. It is true that the human population has little to do with life at the bottom of the pond, but the pond itself is a construction. Scaled up, when we discuss the built environment of towns and cities, we conventionally include intervening green spaces. Beyond the garden lie fields and hills—the product of geology, climate, biota, and time, and so not originally a human construction. However, the flora and fauna is appreciably modified by our choices and decisions. The land is farmed, the range of herbaceous and woody plants greatly reduced and homogenized through grazing by the sheep and cattle deliberately placed in those fields, but also by the numerous rabbits that have invited themselves. In this year, the neighboring farmer has chosen to cut his meadows late in

Figure 10. In a display of environmental modification, termites have constructed a magnificent high-rise residential block and largely cleared the surrounding vegetation. (Source: Paul Lane, with permission.)

Figure 11. Distribution of built and modified environmental elements around a small settlement on Dartmoor, UK. (Source: author.)

the summer to make hay, in contrast to his usual practice of cutting early to make silage. If that change of behavior is maintained for a few more years, we would expect the flora of those fields, and hence their fauna, to change. On the hills that overlook those fields, the August sunshine is opening the flower spikes of heather (*Calluna vulgaris*), which covers many square kilometers of sandstone hills in this part of northern England. This landscape is frequently described as "wild" or even "natural," but it is not.[7] First, this environment is maintained by active human management. Light grazing by sheep removes plant species that might otherwise crowd out the heather, and occasional superficial burning removes old growth and rejuvenates the heather plants. Second, heather is by no means the "natural" vegetation of those hills. There is substantial archaeological evidence of prehistoric settlement and agriculture under the heather, and clear evidence that the soils were formerly deeper and more fertile, supporting a more diverse shrub and herbaceous flora. It was prehistoric agriculture, possibly with a nudge from minor climate change, that pushed the environment of those hills over to the low-diversity flora of heather and bracken fern (*Pteridium aquilinum*), and it is today's management that maintains the heather communities. In short, the purple heather-covered hills are as much a modified environment as the pasture and meadows below them.

In terms of generic structure, then, we need to consider two zones of the human environment: the built environment and the modified environment. Apart from their spatial relationship, there will obviously be some movement of species between them, and the precise boundary may sometimes be hard to define. However, for the purposes of this book, it is a useful distinction to make, separating an environment that is essentially artificial, one that may arise quite suddenly, de novo, from one that, however much it may be amended by human activity, has time depth and perhaps some spatial articulation with environments that are barely modified by people, if at all.

Time is an important parameter in the modified environment. One of the strengths of an archaeologist's perspective is that we see the landscapes around us as a palimpsest of traces of different processes and events in the past.[8] We are accustomed to the idea that we inherit structures such as field systems, burial mounds, or ancient buildings from the past into our present-day landscape, but we must remember that a non-constructed habitat patch—the grassy field within the field system, or the vegetation overgrowing the abbey ruins—also has a time depth that is inherited into the modern landscape. And when we consider communities of animals, such as the birds and mammals that live in the overgrown ruins, they too have a time depth. The attributes of that community today may reflect contingent events in the past as much as they reflect the habitat that we see today. Our present-day observations cannot be other than a slice through a complex bundle of environment and community processes, acting at different rates and with varying degrees of interaction and synergy across habitats that were influenced or modified by, or wholly constructed by, people.

What are the main generic impacts that people make as they build and modify their environment? Obviously, any detailed answer to this question must be highly case-specific. Construction of the Great Pyramid had a quite different environmental impact than the building of the Hanging Gardens of Babylon, which in turn had quite a different impact than laying out New York's Central Park. However, there are certain consequences of human activity, certain repeated generic impacts and outcomes, that can be identified, and that allow us to generalize about the human environment as a place of challenge and opportunity for commensal species. At the risk of oversimplifying the matter, I suggest six major characteristics of the human environment that are particularly significant for other species.

The first, and most obvious, is disturbance. People are enthusiastically active builders and modifiers, and few habitats within the built or modified environment are likely to persist for considerable periods of time without significant disturbance. We are accustomed to the idea of human environments being frequently disturbed, by observing the world in which we live.[9] Urban brownland turns over on a timescale of months to years, seldom decades to centuries. Despite the false impression of stability that the time-averaged archaeological record sometimes conveys,[10] human environments of the ancient past were probably just as subject to disturbance. From an ecological perspective, disturbance (or perturbation, to use the ecologists' rather appealing term) has two aspects: frequency and intensity. Frequency is obviously important, and the timescale of disturbance in human environments varies considerably. Refuse may be cleared daily, roadside verges may be mown fortnightly, a public park pruned and raked annually, a flat roof regraveled every five years. The point is that some disturbance will be on a shorter timescale than the generational span of many of the species that might otherwise colonize that environment, and some will be on a shorter timescale than the seasonal fluctuations of temperature and humidity that will be essential to some species. Small variations in the frequency of disturbance may therefore have substantial outcomes in terms of allowing successful colonization and establishment of stable commensal populations. Intensity of disturbance can also encompass quite a range, from a minor perturbation that temporarily alters the physical structure of a habitat patch to the complete destruction of that patch and all that lived in it. Perturbations that are not catastrophic may temporarily disturb food webs and cause a shuffling of living space just sufficient to facilitate further colonization, giving another species or two a foothold. It is not surprising, therefore, to find that environments that are occasionally and nonintensively disturbed often show a higher biotic richness and species diversity than those that have long-term stability with little perturbation, a phenomenon sometimes packaged as the Intermediate Disturbance Hypothesis.[11] Disturbance, then, is something that characterizes the human environment, and in terms of biodiversity its effects range from beneficial to catastrophic.

A more specific consequence of human environmental modification is the redirection and concentration of surface water. At one end of the scale, this is reflected in the emergence of the great riverine civilizations of Mesopotamia and Egypt. Little could be done about the narrowness of the belt of cultivable land along the Nile, nor could much be done to control the regular inundation of that land. However, the construction of fields and water channels could at least ensure a thorough and consistent separation of land and water between inundations. The same is true even in the quite different setting of the British Isles, where one of the distinguishing features of "natural" landscapes is that they are wet underfoot. People need and use water, but also prefer to stay dry. Surface water, whether in pools or flowing, needs to be managed, moved around, kept in certain places and out of others. We have a rather odd relationship with standing water: appreciating its reflective qualities and incorporating it into our built landscapes, yet aware of it as an inconvenience to movement from place to place and as a possible health hazard.[12] As human modification of the environment intensifies, streams disappear underground into culverts, ponds and damp patches are drained, and bodies of standing water (often with fountains) are constructed where no pond or lake would naturally occur. Those water bodies are then, usually, carefully managed to ensure that only a limited, tolerated range of other species makes use of them: ducks are allowed, leeches are not.

Figure 12. The University of York was built on wet farmland in 1963 and needed the lake in order to manage surface water. Ducks, geese, and other waterfowl find this artificial environment much to their liking. (Source: author.)

That question of which species are tolerated introduces the third significant impact, namely, the replacement of any "undesirable" biota by a "useful/desirable" biota. The subject of human transportation of animals around the world will recur a number of times throughout this book. For now, the crucial point is that colonization of the human environment is not necessarily subject to the many factors that control colonization of habitat patches in general. People decide to remove this species or to introduce that species, and do so with an enthusiasm that can be quite breathtaking.[13] Sometimes this biotic turnover is obviously utilitarian, as when introducing domestic livestock to a newly cleared area of farmland, or extirpating a potential pest species. Ridding a landscape of undesirable animals becomes a worthy act in some cultures, hence the highly doubtful claim that St. Patrick, a holy Welsh expatriate, drove all the snakes out of Ireland. Sometimes the turnover is for aesthetic reasons, as when landed gentry introduced fallow deer (*Dama dama*) to their parks in medieval England,[14] or when someone introduced the attractive but noisy, fecund, and highly gregarious ring-necked parakeet (*Psittacula krameri*) to southern England (chapter 6). Utility aside, the decision that a particular species is or is not desirable will be essentially a cultural one, and it may be susceptible therefore to changes of attitude. One of the most obvious, to which we return in a later chapter, is the cultural rise and fall of the street pigeon. Concerning pigeons, Jerolmack[15] observes that animals become a problem when they transgress spaces designated for human habitation. I think this is too simple. The historical and archaeological records are, as we shall see, full of examples of such transgression being either tolerated or actively initiated. Problematizing the transgression comes later, if at all, and may be due as much to changed perceptions of what is "right" for a place of habitation as to the abundance or attributes of the species concerned. Attitudes to animal neighbors are likely to be replicated across large geographical areas, especially if transmitted by a culturally vigorous and expanding people. The consequences of European expansion across the Old and New Worlds can be seen in the similarity of farm and urban faunas across current or former colonial areas. In some instances this has been relatively harmless (European starlings *Sturnus vulgaris* in the United States), and in some it has been an ecological disaster (rabbits *Oryctolagus cuniculus* in Australia). The term *biotic homogenization* has been applied to this tendency to see the same species recurring in heavily modified environments,[16] in part because certain species have adapted so well to the opportunities that we have offered them, but in part because people have deliberately added and subtracted species according to some culturally informed preconceptions.

A fourth impact, and perhaps one of the most significant, is the human tendency to deposit refuse. The location of ancient settlement sites is usually marked by deposits of food debris, discarded artifacts, and, often, bodily wastes: characteristic elements of Anthropocene sediments. It may be garbage to most people, but it is meat and drink to archaeologists, who eagerly exploit our habitual deposition in order to investigate past lives.[17] We are accustomed to think of garbage disposal and consequent landfill as a modern problem. Certainly the scale has greatly increased within recent decades, just as durable materials such as plastics have become a significant part of the problem. However, the human environment has always been marked by the deposition of refuse,[18] and that propensity of ours has probably been a significant factor in attracting and sustaining commensal animals for millennia.[19] Garbage dumps create a highly modified patch of habitat. If the garbage includes human or animal wastes, there may be local enrichment of nitrogen and phosphorus. That in turn will affect the plant species that grow on the garbage heap, which may in turn affect further colonization by animal species.[20]

Figure 13. A thick cultural sediment under excavation in York, UK, representing the characteristic geology of the Anthropocene epoch. (Source: York Archaeological Trust for Excavation and Research Limited, with permission.)

More important is that a garbage heap constitutes a concentrated patch of food, of potential energy. One way of grasping the importance of this potential energy is to consider the human environment as a discrete ecosystem. That ecosystem can be modeled as energy flows and stores: the trophodynamic model of ecosystems particularly associated with G. E. Hutchinson.[21] In most terrestrial ecosystems, energy enters the system from the sun, captured by green plants and converted into plant tissues. The amount of energy thus captured is the gross primary productivity (GPP) of the ecosystem, and is usually only a few percent of the total solar energy falling on the plants concerned; photosynthesis may underpin life on Earth, but it is surprisingly inefficient. Herbivores eat the plants and convert some of that GPP into herbivore, again with only a few percent efficiency. They, in turn, are eaten by carnivores, which convert an ever-declining share of the original GPP into carnivore. The result is the familiar reduction of available energy as we move up the food chain, and hence the marked reduction in biomass at each successive trophic level. When the set designers for the *Jurassic Park* movies decided to include plenty of fierce carnivorous dinosaurs, they had to include even larger numbers of herbivores, and that in turn required an inconceivably productive flora. (And all of that on an island, furthermore, raising fundamental biogeographic objections!). Narrative impact won out over good ecology in that case, but does not do so in real life. That said, in human environments, people deposit garbage and garbage is food. At the base of the food web, there is a substantial "donation" of energy that is not limited by GPP. That donation will be in forms that will only be available to some species (omnivores, specialist detritus feeders), but those species will draw this additional energy into the food web as they in turn become food for something else.[22] One of the big opportunities for commensal animals comes from that freeing-up of food webs, and as we shall see, some species have seized that opportunity and maintain startlingly high population densities as a result.

The two remaining impacts to be considered here are somewhat connected, as both are a consequence of the ways that we lay out our settlements. From small villages to major cities, people like to have fixed routeways within and between settlement areas. At their simplest, these may only be paths that are repeatedly walked from place to place. At the other extreme, we have the complex communication networks of road, rail, and canal that structure our cities. From footpaths to motorways, these routes can do three things. They constitute a zone of enhanced disturbance, which may in itself be sufficient to discourage certain species from taking up residence nearby and may even become a zone of enhanced mortality.[23] They may become a barrier, either physical or perceived, to the movement of a species, perhaps inhibiting colonization of a new habitat patch, or movement away from an unfavorable disturbance. Conversely, roads, railways, and canals may be valuable corridors, connecting habitat patches within a town, or connecting the urban environment with the terrain around it. Although these "green corridors" are a welcome addition to the urban scene, often hailed as "linear nature reserves," their role as routeways for wildlife may be ambiguous.[24] A study of plants and invertebrates around Birmingham, UK, for example, showed that green corridors make little difference to the dispersal of invertebrate animals, but that they are significant routeways for small mammals.[25] Routeways made by people for people are a significant attribute of the human environment, then, but we must be careful not to overgeneralize about their role in the dispersal and ecology of other species.

Finally, of course, people tend to construct a built environment—the scale, structure, and composition of which will vary considerably, but which will have certain common factors from the point of view of a would-be commensal animal. The towns and cities with which people have speckled the globe obviously vary from continent to continent according to cultural traditions and history, interacting with variables such as climate and available raw materials. Many permutations are possible. An hour spent communing with Google Earth will quickly show contrasting urban scenes, from the densely packed stone terraces of Bradford, UK, to the wide boulevards of Canberra, Australia, to the sprawling *favellas* of Rio de Janeiro, Brazil, to the gleaming pinnacles of New York, USA, and many more besides. However, our concern is with built environments as places for other animals to live. Those other species are not, one presumes, preoccupied by the aesthetics of architecture. An urban cliff is still a cliff, whether it is the sandstone Victorian gothic of London's Natural History Museum or a concrete high-rise in downtown Shanghai. At least some of the apparently bewildering variation in building design and urban layout can be ignored for our present purposes. What matters are the material composition, topography, microclimate, and feeding opportunities presented by the built environment. Le Corbusier's description of a house as "a machine for living in" is curiously apposite.[26]

Today's homes and their agglomerations are mostly built from a few inorganic materials: brick, stone, concrete, and various aggregate blocks, adobe, glass, and steel. Timber and other organics are less often the majority material. Composition alone, then, gives the modern built environment certain characteristics: inorganic, hard, tending to be dry. This last is unsurprising. We prefer our buildings to keep us and their interior dry and to shed water from roofs and surrounding surfaces, often the first step in the redistribution of surface water discussed above. Concrete and structural materials based on it tend to give the built environment another important characteristic, that of being calcareous, derived from the lime of the

Figure 14. Above, a disused railway line has become overgrown and makes a highly successful corridor for wildlife. Below, the same line just a few kilometers away has been converted to a cycle path and is much less amenable to wildlife. (Source: author.)

Figure 15. Kittiwakes *Larus tridactylus* turn a seafront hotel in Scarborough, UK, into a manmade cliff. Note the anti-bird spikes on the ledge to the left of the birds. (Source: author.)

cement component. As concrete weathers, or where rubble occupies a demolition site, there is a colonization opportunity for plants and animals.[27] The various controls on colonization rate and success are complex and may be highly site-specific. What matters here is that colonizing plants and animals need to be able to cope with, or benefit from, a rather dry, calcareous environment. Waste ground in towns in the UK typically grows buddleia, not acid-demanding rhododendrons, and buddleia bushes are well known as feeding stations for butterflies and other nectar-feeding insects. An unusual example of the stochastic nature of the colonization of habitat patches came in summer 2008, when a disused site close to my home, which is always rich in buddleia, grew a fine drift of bee orchids (*Ophrys apifera*) on a patch of mossy concrete rubble.

A fundamental characteristic of buildings is that they are occupied by, and therefore undergo disturbance from, large numbers of people. For many animals, that will render building interiors and ground-level surrounding spaces just too disturbed and hazardous to be satisfactory living spaces. However, people are relatively large animals, so restricted spaces within and below buildings are potential animal habitat. People also tend to avoid the external surfaces of tall buildings, leaving walls and roofs as available space. From the point of view of an urban bird, the modern city is an agglomeration of plateaus on which there is relatively little disturbance, flanked by vertical cliffs that give protection from non-flying predators, and separated by deep canyons at the base of which there is intense disturbance but also locally rich feeding opportunities. Urban pigeons, descendants of the rock dove *Columba livia*, seem to have no difficulty in recognizing buildings as cliffs and ledges, and this perceived "problem" (i.e., of fouling, allergens, pathogens, and ectoparasites) has spawned a substantial literature discussed further in chapter 6.[28] Pitched roofs may be less inviting in themselves, but are generally accompanied by eaves that provide a sheltered wall-top location for cliff-nesting species such as hirundines and swifts. The deliberate engineering of the built environment to provide colonization opportunities for other species is the province of *reconciliation ecology*—beyond the remit of this book, though of some relevance.[29]

Other species to have adapted to urban buildings include seagulls (mews). In Britain and coastal Europe, herring gulls *Larus argentatus* and lesser black-backed gulls *L. fuscus* have found flat roofs to be an excellent alternative to their habit of nesting on beaches and shingle spits. In Venice, yellow-legged gulls *L. michahellis* have increased their breeding success by moving from the islands of the lagoon into the Renaissance conurbation of the city.[30] For peregrine falcons *Falco peregrinus*, the urban cliffs provide not only plenty of perching and nesting space away from human disturbance, but also plenty of prey at densities unlikely to occur outside urban environments. People deposit the garbage that attracts the pigeons, and construct the adjacent cliffs from which to hunt: for a falcon, what's not to like? The recolonisation of central London by peregrine falcons without their deliberate reintroduction has been one of the surprise conservation success stories of recent years. Larger modern buildings provide not only a protected exterior, but an inviting interior world of cavities comprised of suspended floors, service ducts and roof spaces. For a cavity-dwelling domiciliary species such as a house mouse, a modern building is a 3-dimensional network of routeways and living space, protected from the larger predators, with minimal competition, and readily adopted by a largely nocturnal burrowing species. The interior space of a pitched roof, often subject to minimal disturbance from human occupants of the building but benefiting from their warmth in winter, provides nesting space and roosting spaces for birds and bats. The important point to make here is that most of these attributes of the modern built environment are not particularly modern. Artificial cliffs date back to the first buildings to be pushed above single-storey height, in ancient Egypt and Mesopotamia. Pitched timber roofs and their inviting eaves were probably a feature of the buildings of Neolithic northern Europe. It is a nice thought that the earliest farms on the North German Plain, over seven millennia ago, may have had swallows and house martins nesting under their eaves.

In short, although we have looked at the human environment from a modern perspective, many of the attributes that would have presented opportunities to other species are likely to be of some antiquity. Modern towns have degrees of pollution by noise, light, gases, and airborne particulates that are unlikely to have been matched in the past, but these are all negatives, likely to deter other species. The positives—walls, roofs, cavities, warmth, garbage—will have been available for millennia, and it would be naive not to assume that species made use of them.

2

Sources of Evidence

Up to this point, we have considered commensalism rather in theory or principle, as a life strategy adopted by some species. Another way of approaching the topic is to ask how we recognize commensalism: how do we decide that the term applies to this or that population of animals, either today or in the past? This may seem self-evident, but we shall see that there are many often subtly different ways of using human living space for food and shelter, necessitating criteria that can be applied in a wide range of circumstances. Furthermore, our concern here is with the past and present of this strategy, so we need criteria that can enable us to recognize commensal behavior in the past beyond direct observation. Other country that it is, the past presents further challenges. We have contemporary documents and recorded observations for the more recent centuries, but even here we have to take care to understand what people may have chosen to record or not to record, and the motivation or agenda that may have influenced the making of that record.[1] For much of the human past, of course, we have no contemporary documentation, and the archaeological record, the surviving scraps of material evidence, is all that we have. Material evidence—most of it garbage in at least one sense—has biases of its own, though it is perhaps less susceptible to deliberate "storytelling" than are most historical documents. For times such as the medieval period where the two sources of evidence overlap, we should not assume that they will concur, or be concerned if they do not.[2]

THE CONTEMPORARY RECORD

The whole point of drawing attention to our commensal neighbors is that they are the familiar everyday animals that live around us, so we hardly need criteria for recognizing them: we know them when we see them around our homes and gardens or hear them rustling through the attic. Rats, pigeons, and urban foxes are clearly subject matter for this book. Some other familiar species may need more consideration. Starlings *Sturnus vulgaris,* for example, are abundant enough around our towns and cities. Are they commensal animals? The abundance of starlings in some English towns, especially in winter, can be quite striking. The species has an evident taste for Victorian buildings, developing huge roosts on municipal stone buildings and on seaside piers.[3] However, although starlings will take advantage of food handouts in parks and gardens, they are by no means specialist garbage feeders. Many of the birds in those huge winter roosts will fly *out* of the town center by day in order to feed on nearby farmland, or on the grasslands provided by our sports grounds and recreation spaces. This is almost the inverse of the strategy applied by some foxes that live on the leafy edge of town by day, then venture into the urban mean streets by night in order to raid garbage bins. Both species are commensal for the purposes of this study, however. The foxes benefit directly from food and other debris around human settlements, while starlings make direct use of our built environment for shelter while gaining feeding benefit from the adjacent modified environment. By colonizing buildings, starlings are simply adapting, or exapting, a behavior that is typical of the species.[4] In rural areas, starlings develop similarly huge winter roosts in reed beds, swirling like great clouds of chattering smoke before descending to pass the night in the relative warmth and safety that comes with sheer numbers. That same behavior readily adapts to a municipal building or to a seaside pier.

Even present-day observational evidence has to be used with due care. In the Internet age, informal, anecdotal records are quickly spread and proliferated, and it can be remarkably difficult to trace the original factual observation on which one charitably assumes they were originally based. Do urban foxes actually prey on household cats? Or has a fear that they might do so elided into "knowing" that they do? Or did one fox, once, get the better of an unfortunate cat, and therefore it is assumed that "foxes eat cats"? Perhaps all of the above? Similar questions attend the predation of cats by coyotes in North American towns, and we return to this topic in a later chapter. The point here is the care that has to be exercised when using informal records of animals raiding garbage bins, living in roof spaces, or entering houses in search of food. We need to ask: How reliable is the record? Was the behavior repeated by one individual animal? Has the behavior been repeatedly observed for this species? For example, hedgehogs *Erinaceus europaeus* are well-known for eating cat food; canned cat food is generally recommended for fattening up young hedgehogs so that they survive the winter hibernation, and hedgehogs that are accustomed to being fed become quite tame.[5] A couple of years ago, I noticed that my two cats seemed to have developed a greatly increased appetite, and suspected an uninvited guest. Sure enough, one evening I disturbed a hedgehog while it was pushing its way in through the cat flap. How very easily "hedgehogs will eat cat food" could have become "hedgehogs will find a way into your house in order to steal cat food." In fact, the miscreant appeared to be one particularly audacious hedgehog, perhaps one that had become habituated to people when young, and the incident did not represent behavior typical of hedgehogs in general, or of our local population of them in particular. The remarkable and the unusual will proliferate more rapidly than the mundane, so anecdotal sources, especially from the Web, have to be used with real care. On the other hand, such sources are the product of years of informal observation by millions of people, and are valuable for that reason.

What about formal "scientific" data? The research for this book has made considerable use of academic journals from a wide range of fields, especially in behavioral ecology and animal biology. The advantage of these sources is simply the convention in academic publication that supporting data are presented and that sources are identified. Much of what we reliably know about commensal animals stems from published scientific work. That said, the academic literature is not unproblematic. Until relatively recently, the familiar commensal species have been distinctly underrepresented in the mainstream academic literature, at least as a guild worthy of study in their own right. By and large, these species are not threatened or rare, nor are they attractive to grant-giving bodies and therefore to ambitious young biologists. As my conservation-biologist son puts it, "Lemurs are cool, Dad."[6] Of course there are exceptions: the remarkable work on urban foxes undertaken by Stephen Harris and his team at the University of Bristol is rightly renowned, and Stanley Gehrt and colleagues have published extensively on urban coyotes and raccoons in North America.[7] Those exceptions notwithstanding, a lot of the literature on commensal animals either originates in, or is clearly intended to contribute to, the discipline of pest control. For example, whereas avian biologists of an earlier period could write about pigeons in relatively nonjudgmental terms,[8] more recent literature tends to concentrate on their role as vectors of pathogens,[9] or on their breeding behavior in order to discourage it.[10] Commensal animals that are not regarded as a significant pest may be surprisingly understudied. When numbers of house sparrow *Passer domesticus* underwent a rapid fall in the UK recently, it brought to light how poorly this familiar neighbor was known. Fortunately, with the increasing recognition of the educational and therapeutic role of urban wildlife, our commensal neighbors are being regarded as something other than a potential health hazard or too uninteresting to study.

HISTORICAL SOURCES

This is not the place for a full treatment of the representation of animals in historical sources. Linda Kalof's *Looking at Animals in Human History* (2007) is a thorough review. Just a few points need to be made regarding historical sources as evidence.

The first is that the further we go back into the historical past, the more it is the case that surviving documents are a small and by no means random sample of the written records of the time. In archaeology and paleontology, we are accustomed to think in terms of *taphonomy*, the complex series of processes that act between the death of an organism and its eventual recovery as a fossil.[11] Those processes destroy most of the evidence, leaving a surviving residue that is biased and modified in many different ways. In order to understand how, indeed whether, that residue may be used as evidence of any past processes or actions, we have to understand the taphonomy of our assemblage. Something similar applies to historical sources, requiring us to ask, "Why has this source survived, and what was its original intent?" Answering that question fully is the preserve of documentary historians, of which I am not one.

Second, when we seek evidence for the past relations between our species and another, the presence of that species in a historical source may be significant, but its absence is not. The best-known example of this is probably Daniel Defoe's *Journal of the Plague Year*, written in 1722 but referring to the events of 1665. Despite dealing with urban London during one of its most squalid periods, which is saying something, Defoe manages to say nothing about rats. That surely does not indicate that London was temporarily devoid of rats, only that Defoe, for some reason that no doubt made sense to him at the time, saw no reason to include rats in his narrative.[12] Later in the eighteenth century, Gilbert White did take note of rats, observing in 1777 that the "Norway rats destroy all the indigenous ones."[13] This is a useful record, made at the time at which the Norway (or common or brown) rat *Rattus norvegicus* was colonizing England, at the expense of the black or ship rat *Rattus rattus*,[14] and White reflects the observation of a contemporary that the incoming species was "destroying" the other. Today, we would probably think that term inappropriate, as it is unlikely that brown rats directly assaulted and killed black rats. Similarly, we would hesitate to describe black rats as indigenous, which they are not. However, to White, knowing nothing of the archaeological record of this species in England, black rats had "always" been part of the fauna. In a way, Defoe and White reflect the same two categories of sources that I proposed for modern evidence: White, the cautious observer and meticulous recorder, can be seen as an early instance of "scientific" literature, while Defoe, the novelist and political satirist, is more of an anecdotal source. Indeed, the geneticist and writer Sam Berry has argued that the science of ecology developed its very distinctive field-based attributes in Britain, in contrast to Continental Europe, precisely because of the "naturalist" tradition represented by the likes of Gilbert White and John Clare—the latter memorably described by zoologist James Fisher as "the finest poet of Britain's minor naturalists and the finest naturalist of Britain's major poets."[15]

Third, beware of nonsense rehashed from classical sources such as Pliny and Aristotle. Despite the eighteenth-century Enlightenment, the tradition of recycling classical sources uncritically persisted even into nineteenth-century writings on natural history. Bell's 1837 *A History of British Quadrupeds* is a nice example.[16] Bell clearly sets out to be scientific: his accounts of various species are prefaced by Latin names and synonyms based on the Linnaean system, and his species descriptions are quite concise and objective. He is frequently dismissive of his contemporaries (especially French contemporaries such as Buffon) for their reliance on classical sources of dubious merit. Nonetheless, in an otherwise sober and interesting account of the hedgehog, Bell cannot resist mentioning "the use made by the Romans of the prickly skin of the Hedgehog in hackling hemp for the weaving of cloth,"[17] for which no source or supporting evidence is given. Similarly, his account of the two rat species includes a good deal of anecdote interdigitated with sound observation, and he expresses at length some distinctly original ideas concerning the origins of the cat.[18] If Defoe represents historical anecdote and White is early field science, Bell rather splendidly and somewhat arbitrarily combines the two.

Nonetheless, taking those caveats to heart, there is information on our commensal neighbors to be derived from historical sources, and this volume makes use of those sources where appropriate and to the author's limited ability. Some are directly relevant and useful. Stefan Buczacki[19] cites a document of 1356 that refers to people dispersing flocks of pigeons around St Paul's Cathedral in London by throwing stones at them. In fact, the gist of the document is to complain that the stone-throwing Londoners had broken several of the cathedral windows, but it gives us contemporary mention of urban street pigeons in sufficient numbers to prompt a reaction, and a rather negative reaction. Depictions of animals may also serve to put a species in a particular time and place, but even greater care has to be exercised. Dürer's famous 1504 engraving of Adam and Eve includes a cat and a mouse, the tail of which is caught under Adam's foot. The cat looks typically smug and disinterested. What is their significance? Are they simply two familiar species of hearth and home, or are they of some particular symbolic significance?[20] At one extreme, of course, depictions of animals are clearly not intended to be taken literally. The frequent depiction in medieval and later art of unicorns clearly habituated to the presence of maidens, if not of people in general, is obviously mythic, but what about the even more frequent depiction of rabbits in such scenes? Does that reflect the abundance of rabbits in the medieval countryside, or are rabbits symbolic of fertility?

A

HISTORY

OF

BRITISH QUADRUPEDS,

INCLUDING THE CETACEA.

BY

THOMAS BELL, F.R.S. F.L.S. V.P.Z.S.

CORRESPONDING MEMBER OF THE PHILOMATHIC AND NATURAL HISTORY SOCIETIES OF PARIS, AND OF THE ACADEMY OF SCIENCES OF PHILADELPHIA. PROFESSOR OF ZOOLOGY IN KING'S COLLEGE, LONDON.

ILLUSTRATED BY NEARLY 200 WOODCUTS.

LONDON:
JOHN VAN VOORST, 1, PATERNOSTER ROW.
M.DCCC.XXXVII.

Figure 16. Thomas Bells's *A History of British Quadrupeds* of 1837. A fine example of early natural history combined with baseless anecdote.

Figure 17. In Albrecht Dürer's engraving *Adam and Eve* (1504), note the mouse caught under Adam's foot, and the nonchalant, smug cat. The parrot is presumably symbolic.

To sum up, historical sources can give us useful information about the place of animals in the past human environment, such as the flocks of pigeons in fourteenth-century London. In every case, we have to recognize that the surviving document or depiction is a small sample, often being seen out of its context, and usually lacking any reliable guide to its interpretation as literal, objective record, or symbol-laden myth, or simply fantasy.

ARCHAEOLOGICAL SOURCES

The point has already been made that historical and archaeological sources do not give the same record of a particular time and place. Each has its characteristic biases, its strengths and weaknesses. One of the strengths of the archaeological record for the present purposes is its sheer quantity. The archaeological record of other animals consists in the main of animal bones. Most archaeological excavations of whatever date in many parts of the world will yield quantities of animal bones.[21] Excavations on sites of former human settlement, where we might expect to find the remains of former commensal species, are often the most productive of bone fragments. The great majority will be food debris, the remnants of animals that in life may also have yielded milk, eggs, and wool, or may have pulled carts and plows, but finally their carcasses were butchered down to cuts of meat and consumed. With them will be the remains of companion animals, perhaps disposed of with more care than the food debris (the family dog given a "decent" burial in the backyard), and remains of the settlement's own fauna. How often and how reliably the bones of commensal animals will have entered the archaeological record will depend on the nature of their synanthropic relationship. Domiciliary species such as house mice are likely to have lived and died largely within houses and other structures, so their remains would be expected to occur within the excavated structures, drains, and refuse pits of human habitations. Species that utilized the modified environment of the settlement but were not domiciliary are less likely to enter that record, depending on the particular circumstances in which the death of individuals or groups of animals occurred, which in turn would depend upon the contemporary human attitudes and activities. For a number of years, I worked in a city that had a large winter roost of starlings, a memorable sight against the watery sunset. The starlings were noted locally—they entered the documentary record, usually in terms of how to reduce their nuisance value—yet I only rarely saw a dead starling, showing how infrequently that large and conspicuous population would have entered the archaeological record. As with the historical sources, we have to consider the taphonomy of each archaeological assemblage of animal remains. What was the spatial and temporal "catchment" of that assemblage? What period of time and place does it represent, and therefore what range of processes and human activities?[22] Taphonomy and context are crucial. There may be a big difference between cattle bones from a medieval town dump and those from kitchen waste in a household rubbish pit in the same town and of the same date. Those differences largely reflect interactions between people and that other species, which interactions lie at the heart of our understanding of commensal animals.

Animal bones, regrettably, do not come labeled with their species names, and have to be identified. Identification is based in part on first-principles comparative anatomy (this is the tarsometatarsus of a passerine bird . . .) and in part on comparison with modern reference specimens of known identity (. . . and in size and morphological detail it matches jackdaw

Corvus monedula and no other species). Not all excavated bones can be identified. Some may only be characteristic of a higher taxonomic level within which species identification is problematic (radius of a small rodent), or may be too fragmented to show characteristic morphology (rib fragment of a large mammal). One of the many filters that lie between us and the formerly living populations of animals, therefore, is our ability to identify their remains with precision and confidence. New developments in biomolecular science are giving us new means of identifying otherwise undiagnostic fragments by their surviving DNA content,[23] or by sequencing the amino acids in structural proteins such as collagen.[24] DNA-based methods require survival of this rather ephemeral molecule and tend to be too expensive to be used other than for specific cases. Structural proteins are more durable in the archaeological record, but give less precise resolution of species than can be obtained by anatomical morphology. These biomolecular approaches are proving themselves to be a valuable addition to conventional means of identification, but they are unlikely ever to replace morphological identification where that is possible. Identification is a particularly acute challenge with commensal animals, where we may be, in effect, trying to distinguish not merely a particular species, but a particular ecomorph within that species, one that is defined largely by its behavioral adaptations. Commensalism is a phenomenon of the behavioral ecology of populations, not of the anatomy of species.

Figure 18. These bone specimens from an archaeological site in Tanzania are readily identifiable as metapodial elements of artiodactyls (even-toed ungulates), but much more difficult to identify to precisely named antelope species. (Source: author.)

Figure 19. Footprints in a 6,000-year-old land surface exposed on the beach at Formby Point, UK, are another form of evidence. (Source: author.)

Apart from their bones, animals occur in the archaeological record in other forms, albeit much less frequently. In cultures that made bricks and tiles, we may come across the paw prints of animals on structural materials. In the Roman world, for example, the manufacture of clay tiles included a period of time during which the wet clay tiles were allowed to dry to "leather-hard" state prior to firing, and the production of mud bricks in arid climates involves a similar drying stage. During that drying period, animals often had access to the tiles, leaving paw and hoof prints that were subsequently baked into the tile for posterity.[25] Perhaps unsurprisingly, these prints are predominantly of people, including children, dogs, and cats. Animals also leave other traces. An animal will deposit only one skeleton, but during its life will deposit a considerable amount of fecal matter. Much of that will decay and quickly disappear from the record, which is just as well, but occasionally desiccation or mineralization will lead to feces, then known as coprolites, surviving in recoverable form. If the coprolite can be identified to species, it then serves as evidence for the past occurrence of that species in a particular time and place.[26] Furthermore, whereas bones record where an animal was *dead*, coprolites record where the animal was *alive*. That said, coprolites may be difficult to identify with confidence. Human feces may be identifiable at sites where people have lived for some time, and can be an important source of information on diet and health. Other coprolites may be more problematic. For example, rodent coprolites from the site of Alero Mazquiarán, Argentina, yielded ova from intestinal parasites that could be identified to species, but the rodent species involved, which may well have been living commensally with the human residents, remained unidentifiable.[27]

Ancient art and artifact also yield depictions of animals, which may serve to place a species at a particular locus. The point has already been made that many cultures have adopted animal imagery for many different purposes, and the same caveats apply here as apply to art of the historic period. European prehistoric art such as the cave paintings of the Upper Paleolithic tends to represent the large animals that were highly significant in their role as preferred prey. Thus we have many, often evocative and rather wonderful images of mammoths, bison, aurochs, horses, and reindeer. There is much debate over whether these images are naturalistic, objective representations of the animals, or symbolic, mythic, even "dream" representations. Suffice to say that genomic studies show that the horses, at least, are represented naturalistically.[28] More to the point, although the Paleolithic artists depicted hundreds of large animals, they almost entirely ignored the rest of the fauna around them. We shall see that foxes were common enough in Europe at the time, yet they do not appear in the contemporary art. Neither do birds, yet surely there were plenty of them about. The absence of everyday commensal animals in Paleolithic art cannot be taken as evidence of their absence.

Figure 20. Cave lions at Grotte Chauvet are an example of paleolithic animal art at its most evocative. (Source: Don Hitchcock.)

In a quite different cultural setting, the art of Central and South America is full of animals—some of them, such as jaguars, condors, and snakes, clearly derived from the regional fauna. Again, though, this is highly selective representation, centered on animals that had magical significance, ignoring those that did not, yet were surely present—where are the armadillos in Mayan art? Art and artifact can add to the archaeological remnants of animals to show something of the human response to the fauna around them, and the cultural incorporation of some of those species, but it is unlikely to yield any useful information on the everyday commensal animals.

Context is crucial in recognizing past commensal populations. Where have the remains been found relative to the likely past "wild" distribution? That question unpacks into the proverbial can of worms: how can we know the past distribution of species other than by where we have found their remains, and therefore how do we decide that a species found at this site was beyond its "natural" range? A good understanding of the ecology of the species and of past environmental conditions will be important here, as will consideration of geographical barriers that animals are unlikely to have overcome without human assistance, deliberate or not. Occurrences of animals on islands feature repeatedly in later chapters for precisely this reason. Frequency of occurrence in the archaeological record may also be significant if it indicates that a particular species co-occurred with people more frequently than seems likely for a free-living, "wild" population. First-principles ecology requires that herbivores will be far more abundant than carnivores in a conventional GPP-driven food web. Thus, if the archaeological record includes a carnivore species far more often than a similarly sized herbivore, something has acted to "select" that carnivore. Of course that may be human selection of a preferred prey, but that in turn will have depended to some extent on the relative abundance and availability of the species concerned, and so may reflect a situation in which human activities attracted the carnivore in question and/or influenced local food webs to the extent of accommodating an abnormal relative abundance of carnivores and herbivores. Finally, of course, there is what we know of the behavioral ecology of that species, seen in the context of the generalizations that we can make about successful commensal adapters (see Introduction). It is important that our knowledge of the species today does not dictate what we infer about its past distribution and behavior, as some species may have shown a greater range of adaptations in the past than we have seen in them in recent times, but knowledge of the species today at least shows possibilities and potential adaptations, an inclusive but not exclusive repertoire.

There is one further source of evidence that may be helpful, one that is both ancient and modern. The last twenty years has seen DNA analyses develop from an expensive and esoteric science to a widely applied source of evidence on many topics (and a clichéd plot device for crime authors). Genomic studies of modern animals enable investigation of maternal and paternal lines of descent, and the recognition of genetically similar clusters within the diversity of a single species. Those clusters often show some geographical basis, which in turn may reflect the past distribution and movement of the species concerned, including our own.[29] The modern DNA, in other words, documents the history of the species, though reading that document may be quite problematic and often requires comparisons with other sources of data. However, genomic studies may enable us to disentangle past and present populations within a commensal species—for example, among the many commensal rat populations of the Pacific region (chapter 5), or house mice in Taiwan.[30] With the development of rapid DNA-sequencing methods and of vast libraries of sequences for different species, the questions that can be asked of ancient and modern DNA studies increase almost daily. This volume reflects some of that potential, but it is quite clear that ongoing research will quickly render this aspect of the present study out of date.

In sum, all of the sources that we have for our neighbors today and into the past have their strengths and weaknesses, their biases and blind spots. We should not expect them to agree. In fact, the discrepancies between, for example, the archaeological and historical sources may themselves be quite informative. Investigating our commensal neighbors means that we are looking for evidence that is dispersed and often rather biased by perception of nuisance and vermin in the modern literature, only sporadically to be seen in the historical sources, and needing careful, contextually sensitive interpretation in the archaeological record. Perhaps that is why this book was not written years ago!

3

The Archaeology of Commensalism

One of the most intriguing questions regarding our species and our animal neighbors is to wonder how far back into our prehistory such associations extend. To some extent we can address that question through the historical and archaeological records, looking for direct and indirect evidence that may show commensal relationships. Digging into that rich and diverse record to examine the material evidence for a number of different times and places might seem to be the logical place to start. However, such an approach would be open-endedly empirical, looking for evidence without having considered a priori what form that evidence might take. First, then, we need to approach the question from a different direction by asking what environmental impact our species and its ancestors may have had in the more distant past. Where and when might other species have found some adaptive benefit in living more closely with humans, or we with them? Perhaps *hominins* would be better than *humans*, as this exploration of our ecological prehistory needs to begin before the appearance in the fossil record of something that we would recognize as human. Because modern humans have become so effective at constructing the environment around themselves, we may tend to assume that much of what we associate with that modern environment, including our commensal neighbors, is of relatively recent origin. The discussion then becomes circular: if we only look for the attributes of the modern human environment in comparatively recent periods of time, we can be sure of never finding those attributes anywhere, or anywhen, else. By taking a starting point long before the advent of modern humans, we can work forwards in time looking for the earliest traces of familiar human environmental modification, and adaptation to it by other species.

Perhaps, too, we can avoid the mistake of associating "man-made" environmental change only with Western industrialism, seeing it as a damaging byproduct of global capitalism.[1] Certainly our capacity to change the face of the Earth has reached new levels since the end of the eighteenth century, for some the time of the Anthropocene epoch. However, in recent years we have realized more and more that past cultures and societies had their own impacts—perhaps smaller in extent and less in intensity than today, but nonetheless a marked ecological impact.[2] A consequence of our attitude to modern environmental change is our desire to seek out, value, and protect those few and scattered places that we believe (often wrongly) to be remnants of the "natural world." Our prevailing culture draws a contrast between a constructed world of towns, roads, and industry, and a natural world that is unaffected, unspoiled. The origins of this contrast can be traced back to the nineteenth century, to John Muir and other advocates for wild places, supplemented by the 1960s rebirth of environmentalism, reacting against the increasing evidence that pollution was detrimentally affecting large parts of most

Figure 21. This information board on the Isle of Skye, Scotland, shows that the influence of John Muir lives on, but what do we mean by "wild"? (Source: author.)

continents. The fact that modern environmental change is modifying landscapes and biotas that are often already substantially modified by past human activity tends to get lost in the entirely understandable alarm and despondency over recent impacts. Towns, roads, and industry date back at least to classical times, environmental engineering to the riverine civilizations of Egypt and Mesopotamia, and our fractious relationship with fire predates our species.

The aim of this chapter, then, is to look at our evolution as an agent of environmental change, from the earliest emergence of the *hominin clade*,[3] through the introduction of agriculture, to the time of historic towns and empires. What part did close interspecies relationships such as commensalism play in our own evolution, and when and in what ways did people become a significant and attractive source of shelter and food for other species? That would have brought other species into the everyday lives of our ancestors, opening up the possibility that they were culturally incorporated in some way. We tend to regard the other animals whose remains accompany those of our early ancestors as if they were either prey or predator, or irrelevant to the hominin story. That rather assumes that the human disposition to develop affiliative relationships with other animals is exclusively a human, not a hominin, trait—an assumption that is quite unjustified, not least because it is untested.

The earliest part of the story is that of our clade—that is, our species and all of its ancestors and side branches back to our last common ancestor (LCA) with our sister clade, the chimps. The hominin clade can be fairly convincingly traced back through the fossil record for about 6 million years. I say that advisedly, because new evidence emerges from the ground regularly, and usually just as a paper, book chapter, or lecture is put to bed.[4] Modern genetic studies on people and chimps allow us to calculate approximately how long ago our two clades shared a common ancestor, making certain reasonable assumptions about the rates at which genetic mutations arise and gene pools diverge.[5] Such studies lead us to think that our clades diverged about 5–7 million years ago, and therefore that our LCA should be found in the fossil record around that date. But what are we looking for? There has been a regrettable tendency in paleoanthropology to imagine our LCA as something rather ape-like, but modern chimps are as far removed in time from that LCA as we are, so it may have been as different from them as we expect it would be from us. That simple point was grasped by the nineteenth-century biogeographer and evolutionist Alfred Russell Wallace. His observation that "Now, if man has been developed from a lower animal form, we must seek his ancestors not in a direct line between him and any of the apes, but in a line toward a common ancestor to them all"[6] seems to have been ignored for the following fifty years as anthropologists sought to interpret the fossil record as a gradation between apes and humans. Today we try to take a more detached, almost disinterested, view of our place in the evolution of life on Earth, and that view obliges us to proceed with caution. We may expect an early hominin and an early panin (i.e., a chimp ancestor) to look far more alike than do modern humans and chimps, making the recognition of such ancestral fossils extraordinarily difficult. And so it is. Extensive systematic field research in Africa has yielded a modest number of specimens from that critical 5–7 million-year date, and they are the source of heated argument.[7] Those early hominins seem to have been forest-dwelling and quite omnivorous. Perhaps more remarkably, the fossil record indicates some degree of bipedal gait in at least some of them.[8] So those up-from-the-ape cartoon sequences can be filed under "fun but wrong." One suspects that Wallace would have been less surprised by recent findings than some anthropologists seem to have been.

Between about 2.5 and 2 million years ago, our clade became distinctly bushy, with an increasingly rich fossil record apparently representing a number of hominin species, at least some of which probably coexisted. None of the hominins of the earliest Pleistocene could be described as "large-brained"—most had a brain the size of a rather modest grapefruit—but some fossils show the emergence of slightly larger-brained forms with more distinctively "human" anatomical traits in the face and base of the skull.[9] These forms have been named as *Homo* species (e.g., *H. habilis*, *H. rudolfensis*), and are generally assumed to be more directly ancestral to later *Homo* than any other Plio-Pleistocene hominins. Just to complicate the issue, stone tools begin to appear in the record around 2.4 million years ago, and it would be easy

Figure 22. Six million years of human evolution reduced to T-shirt art. The ancestor furthest left is far too chimp-like. The descendant furthest right is far too accurate. (Source: Sonia O'Connor, with permission.)

to assume that the more *Homo*-like hominins were the tool makers. In fact, paleoanthropologists have tended to make exactly that assumption, although the hominins often found in association with early stone and bone tools are the much less "human"-looking *Paranthropus* species.[10] *Paranthropus* is also associated with the earliest evidence, albeit far from incontrovertible, for the management of fire.[11] The important point here is that the years around the turn of the Pleistocene period, about 2 million years ago, saw Africa populated by several closely related hominin species, each with its particular niche and, presumably, distinctive behavior. Although there has been a good deal of speculation regarding the interaction of hominin species with each other, ranging from Robert Ardrey's "killer ape hypothesis" to the opposite extreme of hominins-as-prey,[12] the fact is that we have little direct evidence of how different hominin species interacted with each other or with other animals, beyond evidence that some other species were preyed upon or scavenged by hominins.[13] And by 1.8 million years ago, those African hominins were spreading into Eurasia, becoming involved in new faunal communities, and making themselves a fact of life for a good many other animals.[14]

If there is little direct evidence, then do we have grounds for making some propositions concerning early hominin association with other species? Analogies with extant species are always dangerous when we are trying to understand extinct species, but when it comes to behavior, all we have is the sometimes imperfectly known anatomy and analogy with extant relatives. In fact, one thing that is evident from modern field biology is that associations of different species are not uncommon: anyone who watches birds through the year in a temperate region will realize that small species associate in feeding flocks during the winter.[15] Birds offer some of the best-researched examples. In the North Sea, mixed flocks of gulls (Laridae) and auks (Alcidae) commonly associate where concentrated shoals of small fish offer a feeding opportunity.[16] The auks largely feed by diving, pursuing fish underwater, and tending to drive parts of the shoal towards the surface. The gulls, usually kittiwakes *Larus tridactylus*, feed from the surface, benefiting from the activity of the auks. This seems to be a genuine mutualism, with all species concerned benefiting. Interestingly, these multi-species feeding associations tend to break up when large numbers of herring gulls *Larus argentatus* become involved, as these large and boisterous birds are kleptoparasitic or, to put it bluntly, "scroungers." When one species does not play the mutualistic game, the benefits for all collapse. One of the advantages identified in the North Sea study was that feeding associations allow all of the species involved to gain maximum benefit from each feeding opportunity, and this is particularly important when, as with shoals of small fish, the food resource may be highly concentrated into patches that are dispersed over large areas. Here our imperfect knowledge of Early Pleistocene environments becomes critical. How patchy were they? If mutualism is favored when resources are concentrated in patches, then periods of climate fluctuation, leading to fragmentation of environments, may have made mutualistic or commensal behavior highly adaptive. This would have been especially so in mountainous or riverine regions where topography and drainage would further subdivide patchy environments. The Rift Valley region of East Africa has produced many of the fossils on which our knowledge of early hominins is based. This is a tectonically dynamic landscape with abrupt changes of topography and hence of environment. In resource terms, the Rift must have been a distinctly patchy place to live: productive in terms of environmental diversity, but requiring behavioral adaptations to make the best use of patchy, concentrated resources.

Like birds, primates incline to multi-species associations, feeding together and often reacting to each other's alarm calls. There are obvious advantages in terms of predator vigilance:

field research has shown that individual monkeys feeding in mixed-species associations look up from their feeding activity, presumably to watch out for predators, much less frequently than they do when feeding in single-species groups.[17] Reduced rate of look-up presumably increases feeding efficiency, but one wonders whether the effect is really of sufficient benefit wholly to account for the regularity with which different monkey species associate. Monkeys are also the non-beneficiaries of more straightforward commensal associations. In arid regions of southern Africa, rock kestrels *Falco rupicolus* associate with groups of chacma baboons *Papio ursinus* as the baboons forage for small prey items.[18] Some insects successfully evade baboon fingers by taking flight, and are preyed upon by the kestrels. This is a distinctly asymmetrical relationship in which the kestrels benefit quite directly, neither to the benefit nor the detriment of the baboons, and hence a (+, 0) relationship characteristic of commensalism. Similarly, sika deer *Cervus nippon* in Japan have learned to associate with macaque monkeys *Macaca fuscata* while the monkeys are feeding in trees.[19] The macaques drop fruit and other debris from their feeding, providing the deer with a more diverse range of resources accessible to them at ground level. The monkeys appear to be indifferent to the presence of the deer, though observers have noted more aggressive behavior between deer that are feeding commensally with macaques than among deer feeding as a single-species herd. Changes in intraspecies antagonism seem to be a common observation among commensal animals, and we return to this topic in later chapters. Bushbuck *Tragelaphus scripta* associate in much the same way with groups of baboons feeding in trees and scrub.[20] In short, commensal feeding relationships are common among other species, and a number of animals have learned to take advantage of the feeding behavior of groups of monkeys. Although we can only speculate, it seems highly likely that commensal relationships would have developed between early hominins and other species, and possibly between species of hominin.

Apart from comparisons with other species and argument from first principles, what other evidence can be adduced to investigate the place of commensalism in the human past? In order for close association with people to have been advantageous for another species, particular requirements need to have been met. Human or hominin populations need to have been creating patches of sufficiently modified environment to comprise distinctive habitat patches or to offer the potential of a new niche. Living space and feeding opportunities are the two key parameters, though human use of fire for cooking and warmth and, even if inadvertently at first, a means of clearing vegetation, must have been significant to other species. The first controlled use of fire by our ancestors is of considerable cultural and ecological significance, yet remarkably difficult to trace. Ash is quite fugitive in the geoarchaeological record, though charred bones and modifications of other materials by heat may persist. That only leaves the problem of separating purposive, controlled fire from opportunistic use of natural brush fires initiated, most likely, by lightning. Thus claims for the early use of fire by early hominins at Swartkrans in South Africa[21] have been met with caution,[22] and by some, rejection.[23] The identification of purposive fires requires several different lines of evidence, carefully applied.[24] In the Middle East, a cautious approach shows regular use of fires, principally in caves, by about 300,000 years ago, and John Gowlett notes that by this time fire seems to have been in regular use across much of Europe.[25] The environmental impact of this technological development is generally discussed in terms of foraging efficiency and vegetation changes. We should not forget, though, that a well-lit cave mouth may have been attractive to other species seeking warmth or shelter from fire-averse predators, and the debris from cooked meat and bone may have been more accessible than the raw material to some scavengers. If so, speculative though this is, Europe's Neanderthals may have been more attractive to synanthropic, even commensal neighbors than has generally been allowed.

The possibility that some animals may have adapted to some degree of commensalism with people during the Paleolithic has not up to now been discussed in general terms, and is not easy to address with any rigor through the archaeological record. Continental Europe has quite a rich record for the Middle and Upper Paleolithic, and is probably a good place to start. The overwhelming majority of sites are in caves or rock shelters. This may be in part because people utilized these natural shelters in preference to living in the open or constructing something artificial. However, any survival at all of Paleolithic archaeology other than in caves and rock shelters is rare indeed, so the apparent attraction to such locations may be a consequence of survival of evidence, not evidence of a preference. This inevitably placed people in close association, or in competition, with other species that utilized caves as den or hibernation sites, notably hyena, wolf, lion, and bear species.[26] One of the intriguing challenges of Paleolithic cave archaeology is distinguishing bone assemblages accumulated by people from those accumulated wholly or in large part by other species.[27] A further complication is added by a tendency on the part of those who investigate such sites to make separate studies of the larger mammals that are likely to have been human prey, and of smaller mammals and birds that are thought to be more indicative of the wider environment. Fortunately, this separate publication is becoming less common than it was a generation ago. Apart from splitting up the evidence from the site in question, it means that any interpretation of human involvement with other species is itself taxonomically constrained.

Despite enough caveats to deter all but the foolhardy, an overview of the European Middle and Upper Paleolithic sites shows certain repeated observations that may be relevant here. A valuable study of mammal remains from this period was undertaken by the meticulous John Stewart, with the aim of detecting competition between Neanderthals and modern humans.[28] Stewart's research thus concentrated on the larger mammals, in particular the ungulates, but his data show just how regularly red foxes occur at these sites: present at 123 (out of 239) Upper Paleolithic sites and 33 (out of 83) Middle Paleolithic sites. For comparison, the similarly sized wildcat occurs in 17 and 15 of the sites, and badger in 21 and 6.[29] Two caves in the Ach Valley of southwest Germany typify the detailed occurrence of foxes.[30] At Kogelstein, Neanderthals and hyenas both appear to have been using the cave, though presumably not at the same time. Fox bones are abundant right through the sequence, though identification questions mean that the published data are not differentiated between red fox *Vulpes vulpes* and arctic fox *Vulpes lagopus*. At Geißenklösterle, foxes are present all through the Middle and Upper Paleolithic deposits in bone assemblages that are otherwise mostly of large ungulates. A similar frequency of foxes is seen in Middle Paleolithic sites in the eastern Italian Alps.[31] Evidence of the presence of foxes and other carnivores around human living sites does not always take the form of their physical remains. At the Middle Paleolithic site of Mujina Pećina, in Croatia, carnivore bones were relatively infrequent, but evidence of their scavenging and gnawing of bones discarded by the human residents was quite common.[32] The authors observe, "Carnivores were also active at the site; they visited it, probably following on human occupations, to scavenge bones and other food waste."[33] At the remove of several tens of millennia, and given the complexity of cave stratigraphy, it would be near-impossible to be more explicit about the relationship between people and foxes during the Pleistocene. However, the fact that the species occurs so regularly in places where people (and sometimes other predators) were generating concentrations of food debris is hardly surprising given the habits of foxes. If people and foxes were coming into regular contact in the Paleolithic, that raises the obvious question of how far foxes adapted to exploiting the

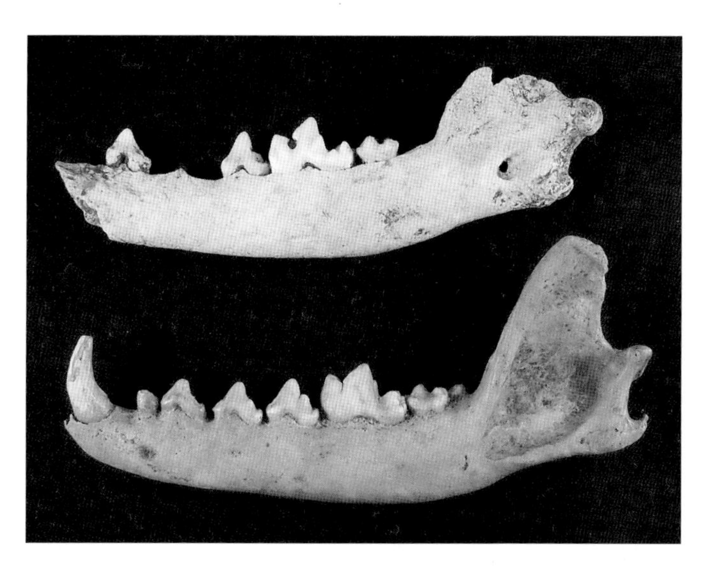

Figure 23. Fox *Vulpes vulpes* bones from an Upper Paleolithic assemblage from northern England. (Source: author.)

opportunity presented by people (and apparently foxes did not discriminate between modern humans and Neanderthals), and how their human contemporaries regarded them: with nonchalant disinterest, or as a significant part of the everyday landscape?

Another group of animals that recurs in the European Paleolithic record are the corvids: the crows, ravens, choughs, magpies, and jays. We meet this intelligent family again in a later chapter; for now, their regular occurrence parallels that of foxes. Jackdaw *Corvus monedula* is often the most common species that is not obviously a "game" bird taken for food, for example at Middle and Upper Paleolithic sites in Central Poland.[34] Elsewhere, the two European chough species *Pyrrhocorax pyrrhocorax* and *P. graculus* occur quite frequently, for example at the important Neanderthal site of Gorham's Cave, Gibraltar,[35] and at sites in the Italian Alps.[36] More remarkable is the large number of ravens *Corvus corax* from the Upper Paleolithic (Gravettian) site at Pavlov I, in the Czech Republic.[37] Pavlov is a famously large site, sometimes referred to as a "megasite," occupied between roughly 27,000 to 25,000 years ago. The mammal bones from this site include abundant carnivores, mostly wolf and the two fox species, and the extensive bird fauna consists largely of game birds of the grouse family Tetraonidae and ravens. Corvids as a whole constitute 46 percent of all identified bird bones, and ravens are by far the most abundant corvids, with just a few bones of jackdaws and choughs. Why so many ravens? In the absence of charring or cut marks on raven bones, it is difficult to be sure that they were eaten. Zbigniew Bocheński and his colleagues point out the large scale of the occupation site and suggest that ravens were attracted by feeding opportunities presented by carcass debris. In the harsh, open terrain of central Europe, such a concentration of potential food must have been attractive. But why were so many ravens not only present, but dead and deposited at this site? Perhaps they were not content with debris and were stealing food, or perhaps their bones and feathers were useful as raw materials. Pavlov has yielded a number of examples of large bird bones worked into tubes for some purpose. Whatever use people made of the ravens after their death, Pavlov seems to be an example of corvids behaving commensally over 25,000 years ago, helping to make the point that it is the opportunity that matters, not the cultural affinities of the human population. Ravens feature in North America, too. At the PaleoIndian site of Charlie Lake Cave, British Columbia, Jonathan Driver excavated two articulated raven skeletons from the space within which people were living and depositing refuse.[38] The fact that the ravens were

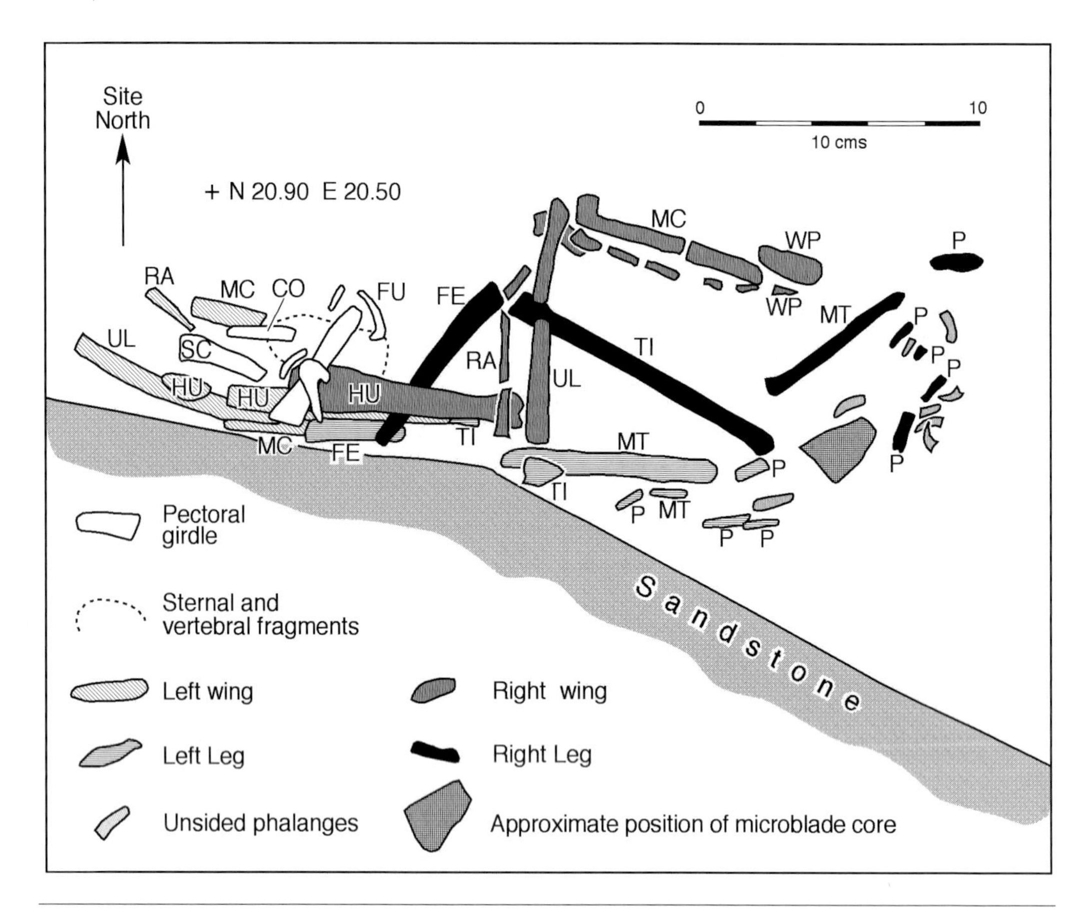

Figure 24. Articulated raven skeleton at the Paleoindian site of Charlie Lake Cave, BC. (Source: Reproduced from *American Antiquity* 64 (2) (1999): 289–98, with permission.)

articulated, unlike other bones in the same deposit, indicated a particular deposition process, and Driver postulates that the ravens lived opportunistically around the human settlement.

Many of the Paleolithic sites discussed so far include remains of wolf *Canis lupus*, a species that must have been a close competitor with humans in the predation of ungulate herds. However, where two species share living space and resources, competitive success of one species over the other is only one of a number of possible outcomes.[39] My own work on bones from Late Pleistocene cave sites in northern England has shown that wolves used caves and rock shelters as dens, and wolf tooth marks are common on animal bones excavated from such sites.[40] Particularly intriguing are specimens that carry tooth marks *and* marks left by stone tools. Where these marks co-occur, we can sometimes see that the tooth marks overlie the cut marks, showing that wolves scavenged bones or carcasses that people had already butchered. This raises an interesting alternative to outright competition: wolves may have gained some benefit from scavenging the debris of kills made by people. Although the evidence to date is slight, there is some indication that bones scavenged by wolves from human kills are from large-bodied animals such as horses and aurochs (wild cattle, *Bos primigenius*). Hunting these big ungulates may have been a serious challenge for wolves, with a low success rate, so it may have been better for them to loosely associate with bands of human hunters in order

Figure 25. Flint-tool cut marks overlain by wolf tooth marks on an aurochs *Bos primigenius* femur from Upper Paleolithic deposits in Victoria Cave, England, show that wolves scavenged a carcass originally butchered by people. (Source: author.)

to scavenge from kill sites, and to concentrate their own hunting on smaller-bodied deer. By scavenging human kill sites, the wolves gained access to prey species that would otherwise have been beyond their capability. Although this is not quite commensalism, it is a more complex relationship than simple competition between predators. Dogs, the domesticated form of the Eurasian grey wolf, are the earliest species that is recognized to have entered a domesticatory relationship with people.[41] Where wolf populations developed the behavioral trait of scavenging from human kills, it is not difficult to see how readily a more complete integration of wolves into the human "pack" may have occurred.

To another species, what really matters about human populations is the extent to which we modify the environment around us, amending and removing some habitats while creating others. As archaeologists, we tend to see humans in the Paleolithic as *affected by* the environment, not *effecting change* in it. However, as we have seen, where people generated scavenging opportunities through their actions as predators or through deposition of residential garbage, even Paleolithic peoples may have created conditions in which a degree of commensalism would confer advantages. As we move into the Holocene, away from the climate challenges of the Pleistocene, human environmental intervention becomes more and more apparent as hunting and foraging gave way to settled agriculture.

Perhaps the best known of the centers in which farming emerged in prehistory is the Middle East. Framed by the eastern Mediterranean, the plateau of Anatolia to the north, the mountainous west of present-day Iran, and the arid deserts of Arabia, the diverse topography of this region must have lent itself to the development of a mosaic of habitats in the millennia before so much of it was homogenized by agriculture and overgrazing. The major rivers—Jordan, Orontes, Euphrates—provide alluvial plains and relatively humid habitats, while plateaus and interfluves give a complex landscape of slopes and higher ground. In the rapidly changing environment of the Late Pleistocene and early Holocene, fragmentation of species' distributions and the local development of a temporary biota out of equilibrium with the regional climate may have been quite common. The term "Fertile Crescent" is often applied to the Jordan-Euphrates arc, but "fertile" is not the same as "stable."

The most intensively studied part of the Middle East is the Levant, the western arm of the Fertile Crescent between the Mediterranean and the North Arabian desert. Here, the climatic fluctuations of the Late Pleistocene coincided with a cultural landscape of specialized hunters, a period known generically as the Epipaleolithic.[42] The Epipaleolithic is usually divided into two quite distinct stages. Epipaleolithic 1, lasting from about the height of the last climate cold stage at 20,000 years ago to about 12,700 years ago,[43] is typified by very small sites of scattered flint tools and bones that probably represent short-lived occupation by mobile hunters and gatherers, assigned to the Kebaran culture. Crucially, sites of this period show no sign of structures, no indication of a built environment on even a small scale. Epipaleolithic 2 (or Late Epipaleolithic) sites represent the period from about 12,700 to 11,400 years ago, and are particularly well-known from sites along the Jordan valley, in the hills of Judea, and around Mount Carmel, overlooking the Mediterranean. In this region, the Epipaleolithic 2 culture is known as the Natufian, after Wadi an-Natuf, where Dorothy Garrod's excavations in the 1920s first defined the culture.[44] Although some Natufian sites are only small and probably only short-lived, this period sees the first evidence for what could be called villages, clusters of substantial stone-built structures, some with paved floors, and a rich and diverse material culture of bone tools and ground stone bowls, mortars, querns, and figurines, including animal forms.[45] Other Natufian sites utilize caves and rock shelters, but show a similar intensity of occupation and deposition. A sense of permanence is given by the presence of human burials at some sites, including a few in which a human burial is accompanied by an animal.[46] The presence of sickles, mortars, and querns indicates the consumption of seeds and similar plant foods in need of reaping and grinding, and it is tempting to see this as evidence for the beginnings of crop agriculture. However, although wild cereals such as barley may have played a part in the Natufian diet, the people seem to have been hunter-gatherers, with an economy based on wild plants and hunted game, especially gazelle *Gazella* spp. The Natufian thus presents an excellent case study: patches of intensively modified, constructed environment, but an absence of "fields."

The Natufian stage—sedentary hunter-gatherers—is often seen as a step on the way to agriculture, and it is probably a reasonable assumption that more or less permanent settlement both facilitated the management of crops and livestock, and necessitated the intensive production that agriculture allows. Peter Wilson makes a more intriguing point: that the development of architecture as a technology was a necessary precondition for the eventual development of farming.[47] Noting that built structures generally predate agriculture, Wilson argues that this was the really significant technological advance, requiring a practical knowledge of surveying and the properties of structural materials. Furthermore, architecture imposed on its human population a much more structured way of living, directed by walls and doorways and paths. So was this also the technological change that most mattered to would-be commensal species? Probably not: we have already possible associations of people, foxes, and corvids arising in the Upper Paleolithic, in the absence of "buildings," and it seems more likely that it was local concentrations of potential food rather than their relatively modest structures that attracted commensal animals to Natufian sites: ecology, not architecture.

The biology of Natufian sites was highlighted by the Israeli zoologist Eitan Tchernov.[48] He noted the sudden and abundant appearance of house mice *Mus* spp. and house sparrow *Passer domesticus* at a number of Natufian sites, in sharp contrast with the earlier Epipaleolithic 1. At the famous site of Hayonim Cave, for example, sparrows constituted over 50 percent of the bird bones.[49] Tchernov argued that this was commensalism plain and simple, but got into something of a circular argument over the association between this commensal relationship and sedentary settlement on the part of Natufian people. If part of the case for arguing that the mice and sparrows were present as commensal animals relies on there being a sedentary human population, is it valid to use the presence of the mice and sparrows to argue that Natufian settlement was therefore sedentary? Tchernov's work was criticized on these and other grounds,[50] though he subsequently refuted the criticisms in detail and at some length.[51] More recently, a detailed examination of small vertebrates from the Natufian cave site at El Wad has shown that although much of the assemblage can probably be attributed to the activities of owls cohabiting with people at the cave, a separate component of commensal mice can be detected.[52]

The important point is that if we accept the capacity of other species to respond to opportunities presented by human activities and structures, then the advent of "villages" and intensively occupied caves in the Natufian could be predicted to have been attractive to a limited number of behaviorally and ecologically preadapted species. Such locations would have offered reduced predation or competition, with a concentrated and predictable food supply, and protected roosting and breeding sites. For relatively small omnivorous, social rodents and birds, the opportunity must have been obvious, so the abrupt appearance of mice and sparrows on Natufian sites should not be a surprise and need not tell us much about the activities of the human population beyond the provision of the habitat parameters outlined above. Regular movements of people between larger and smaller Natufian sites may have made little difference to the commensal animals—especially if food was being stored, giving a food supply that was available in the intermittent absence of people. One further point of interest that Tchernov raised was the possibility that species may have been commensal in the past to a greater extent than they are today. Some Natufian sites such as Raqefet Cave[53] have yielded surprising numbers of spiny mouse *Acomys cahirinus*, a species that is commensal in some North African populations today, though free-living over much of its range.[54] Spiny mouse seems to co-occur at some Natufian sites with the more obviously commensal rats and mice, perhaps indicating the same behavioral flexibility that it shows today.

The presence of dogs in Natufian burials at Ain Mallaha (also known as 'Eynan) and Hayonim Terrace is intriguing, and has been the source of much debate concerning the origin of domestic dogs.[55] Such intimate association of people with a canid would seem to necessitate referring to the animals as "dogs" rather than "wolves," though there are also morphological grounds for doing so. The Natufian dogs are generally rather smaller than wolves from this region, and show a characteristic shortening of the snout and reduction in size of the teeth.[56] The archaeological evidence shows wolves coming into domestication and becoming dogs in this part of the Middle East, and this is consistent with evidence from analyses of dog and wolf genomes.[57] The interesting question is not so much whether this association happened, which clearly it did, but why. A conventional explanation of the domestication of dogs by hunter-gatherer peoples points out that both are pack animals, social predators that are likely to have been competing for the same prey.[58] One can see the advantages for people of having a keen-eyed, fast-running associate on the hunt, but what might have been the advantage for the wolves? Perhaps Natufian settlements and later Paleolithic sites elsewhere in Eurasia offered

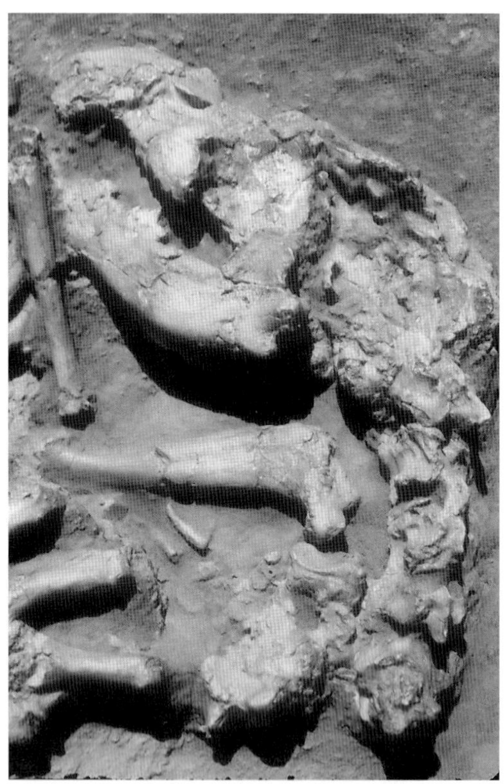

Figure 26. The burial of a woman and a puppy from the Natufian cemetery at Ain Mallaha, northern Israel; the inset shows the woman's hand bones, to the lower left, overlying the puppy bones. (Source: Simon Davis, with permission.)

an opportunity to some of the wolves in a pack, rather than to the species as a whole. Wolves exhibit an agonistic-dominance hierarchy, maintaining a rank order within the pack through aggressive behavior from high-ranking animals towards lower-ranking ones, and something of the same is seen in free-ranging dogs.[59] Rank predicts access to food, such that low-ranking individuals get less access to food, and body size to a large extent predicts rank, such that larger individuals are likely to have a higher rank than smaller ones. It follows, then, that smaller individuals within a wolf pack would have most to gain from opportunistically exploiting alternative sources of food.[60] A degree of commensalism with Natufian settlements may have been advantageous to smaller wolves, bringing them into more regular contact with people and thus encouraging the behavioral changes that facilitated domestication, and constituting a selection mechanism for small body size acting on the earliest populations of domestic dogs.

Around 11,000 years ago, a shift in economic strategy, accompanied by changes in settlement construction and artifact suites, shows the transition to the Neolithic. Exactly what triggered this change has been the subject of much debate. Various models have sought to integrate an often imperfect knowledge of past climate and environmental change with assumptions regarding population growth, mediated through anthropological theories that were sometimes only briefly current.[61] Whether today's consensus is any better may be questionable, though it is based on more, and often better quality, evidence. The Natufian and other entities of Epipaleolithic 2 coincide with the end of the last climate cold stage, in particular with the cold, dry millennium of the Younger Dryas. The replacement of Epipaleolithic cultures by the Neolithic took place as the Younger Dryas climate quite abruptly became warmer and damper in the Middle East. It may be too simplistic to see this just in terms of a shift to more favorable growing conditions; changes in the density and distribution of human populations may have played a role as well. The immediate effect is likely to have been to put biotas into disequilibrium, producing unexpected and ultimately unstable combinations of species in patches of habitat that were themselves unstable. Maybe the emergence of agriculture was an attempt to stabilize shifting plant communities. The post–Younger Dryas warming may have been less dramatic in scale and speed in the Middle East than it was in, for example, the British Isles, where a change from Arctic tundra to present-day temperatures occupied less than a lifetime.[62] Nonetheless, that climate change seems to have been enough to have triggered a widespread change in subsistence strategy in the Levant, and to usher in the Neolithic.

The earliest stages are known as Pre-Pottery Neolithic A and B (PPNA, PPNB), following the type sequence established at Jericho, though the terms Neolithic 1 and 2, and Aceramic Neolithic are also commonly used.[63] The important point is that between 10,000 and about 8,000 years ago, economic strategies in the Levant underwent a gradual transition from growing some field crops alongside foraged wild plants and hunted game, to largely cultivated plants but hunted game, to mixed farming of crops and livestock with only a little foraging and hunting.[64] Neolithic sites are typically clusters of mud-brick houses up to a hectare in extent, though a few are much larger: Abu Hureyra extended to 11.5 hectares of mud-brick Neolithic settlement. In the Neolithic, then, the patches of built environment increased in number and size; cultivated fields of crops, presumably more or less in monocultures, modified the surrounding landscape; and herds of sheep, goats, and cattle effected further changes through their grazing pressure and dung. To what extent Neolithic farmers extirpated, or at least chased away, potential predators of their livestock is not known from direct evidence, though it seems highly likely.

Figure 27. Excavations at Çatal Hüyük, Turkey, show thick "midden" deposits built up around a Neolithic house. It is one of the earliest human settlements to have been colonized by house mice. (Source: Lisa-Marie Shillito, with permission.)

In terms of commensal animals, it is questionable whether the advent of the Neolithic, though sometimes seen as a revolution in human terms,[65] made much of a difference. We have seen that a commensal community of rats, mice, and sparrows quickly adopted the niche offered by Natufian settlements. It is no surprise that this association persisted through the Neolithic, with little indication that the adoption of crop and livestock husbandry affected the commensal niche to any significant extent. This suggests that it was the presence of built structures and the consequent spatial concentration of stored food and occupation debris that was essential, rather than any specific associated activity. From this observation stem two important further inferences. First, commensalism generally predates domestication as a human-animal relationship. And second, Tchernov was right to propose an appreciable degree of sedentism in the Natufian, as the continuity of the commensal niche indicates a high degree of continuity in the human lifeway from Natufian to PPNA.

Another species of note in this context is the red fox. We have already noted an association between people and foxes in the European Upper Paleolithic, and foxes occur frequently and sometimes abundantly on Natufian and PPNA sites throughout the Levant.[66] These foxes are reported and described in varying levels of detail, perhaps because they are not gazelles or caprines and so not the economic mainstay of the site. At El-Wad, close examination of fox remains revealed cut marks showing that some, at least, had been butchered, presumably for meat. This could be taken to show that foxes were hunted and trapped in the same way as the various other medium-sized mammals in these assemblages. However, foxes often outnumber hares, quite the reverse of what would be expected if the two species were being hunted together. At Tel Tif 'Dan, a PPNB site in southern Jordan, for example, fox bones outnumber hare by nearly three to one.[67] The implication here and at similar sites is that foxes were captured separately from other medium-sized mammals, and in greater numbers. This is consistent with a commensal model in which foxes were attracted to the early village settlements by the availability of food from occupation debris and perhaps small rodents, and were captured and used for meat and fur. If we accept that association, the exceptions become interesting. The Aceramic Neolithic site of Körtik Tepe in southeastern Turkey, for example, yielded only a few specimens of fox,[68] whereas foxes were abundant at the slightly earlier site of Hallan Çemi Tepesi in the same region.[69] The difference in date may be significant: foxes were also scarce at the Late Neolithic Anatolian site of Domuz Tepe.[70] Reuven Yeshurun and colleagues[71] noted a decline in foxes in the later PPNB sites in their survey of thirty-one assemblages from the Levant, though a site-by-site analysis shows there to be exceptions to this trend. The site of 'Ain Jamman, like Tel Tif 'Dan, a late PPNB site in southern Jordan, yielded appreciably more foxes than hares,[72] whereas Nevah Çori, also PPNB but on a Euphrates tributary in southern Turkey, shows foxes declining through the PPNB but always outnumbered by hares. Putting all of this together, the archaeological record suggests that foxes were early adopters of the commensal niche in the Middle East, and their reliable proximity to settlements led to their regular capture and use. In time, something changed, whether in terms of fox behavior or human response to foxes, leading to a reduction in the frequency and abundance of foxes in archaeological assemblages from at least some parts of the Middle East.

One possibility to consider is that commensal behavior offered a buffer in the face of the climate changes that marked the Pleistocene-Holocene boundary. In the cool, dry period of the Younger Dryas, biotic productivity may have been reduced sufficiently for an opportunistic omnivore to gain benefit by scavenging around human settlements, and perhaps preying on the small rodents that were taking the same opportunity. That benefit continued into the Holocene, the period of PPNA cultures, a period of environmental change and disequilibrium,

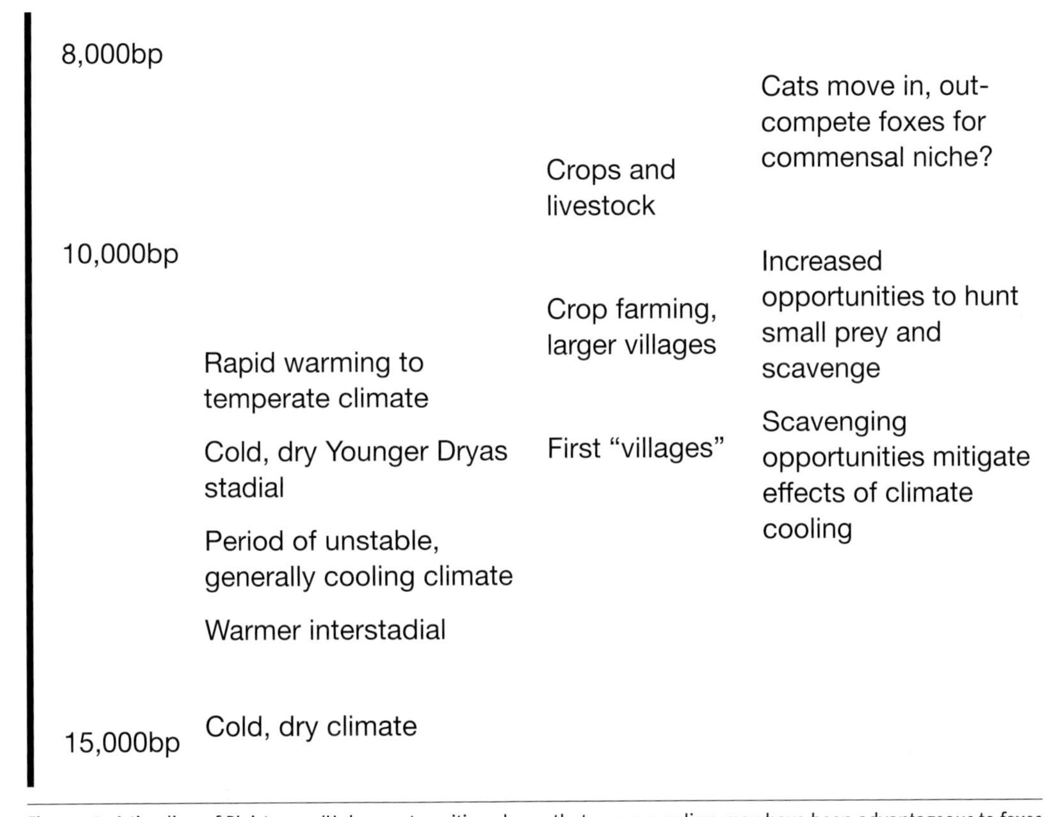

Figure 28. A timeline of Pleistocene/Holocene transition shows that commensalism may have been advantageous to foxes during a period of climatic instability.

only declining later in the Neolithic as conditions stabilized. Another possibility to consider is that foxes may have reduced their association with human settlements in the later Neolithic in the face of competition from another commensal species: cats. There is insufficient published evidence to pursue this point further than speculation, but we return to the place of cats and foxes in commensal communities in the next chapter.

The Epipaleolithic and Neolithic periods in the Middle East thus saw significant changes in people's impact on their environment and in their relationships with other species. Most obviously, several ungulate species came into domesticatory relationships with people, adapting to a "preferred predator" in return for reduction of competition and predation, and a considerable extension of geographical range. Even before that change was under way, other species had constructed a niche within human settlements, adapting rapidly as people adapted to a more sedentary lifestyle in a constructed home environment. As the climatic oscillations of the end of the Pleistocene gave way to the comparatively mild stability of the Holocene, a combination of regional environmental change, human population numbers and density, and simple opportunity led communities in several parts of the world to develop a changed relationship with the biota around them.[73] This change is most obvious in the Middle East, but we should not forget that in other temperate regions of high relief, people were responding to challenge and opportunity in similar ways. In northern Peru, the cultivation of squashes and beans was getting under way by 8,000 years ago,[74] and maize and squash were under cultivation in Mexico by that time.[75] In the great river valleys of China, there are signs of rice agriculture replacing foraging by 10,000 years ago.[76] Most remarkable, perhaps, is the evidence that around 10,000 years ago, people in the highlands of central New Guinea were beginning to cultivate taro around the edges of wetland areas.[77] This is not the place to discuss the origins of agriculture in detail. Suffice to say that in the first few millennia of the Holocene, various forms of farming arose in many places around the world, not by diffusion of ideas from one point of origin, but because the circumstances of the time made it at least advantageous if not an absolute necessity to replace what were often quite complex and sophisticated foraging strategies by plant cultivation and animal domestication. In the Middle East, China, and Mesoamerica, differing degrees of nucleation of population resulted, producing patches of almost wholly artificial environment that we may call "towns," often with highly modified surroundings in which the soil and biota had been given over almost entirely to crops and grazing land. Meanwhile in New Guinea, "gardening" within the highland forests gave way to the construction of cultivation mounds by 7,000 years ago, and complex systems of irrigation ditches before 4,000 years ago.[78] Different though they certainly were, the cultivation strategies applied at Kuk Swamp and in the Euphrates valley had in common that they presented their contemporary faunas with an acute challenge and a range of new opportunities.

If we revisit the list of environmental impacts typical of human settlement (chapter 1), we can see that the advent of farming introduced most of them to some degree. Prior to farming, the hunter-gatherers of the Paleolithic checked off only one impact on the list—refuse deposition—with the occasional hesitant check mark for "disturbance" where people repeatedly aggregated in significant numbers or burned off areas of vegetation, whether by accident or design. From the early Holocene onwards, we can see the other impacts—disturbance, redirection of surface water, replacement of biota, concentration of food sources, construction of routeways—to a greater or lesser degree. Agriculture in many, though not all, parts of the world is based on large-seeded annual plants such as barley, wheat, rice, maize, and millet. The impact of agriculture is to replace what were probably dispersed stands of these plants mixed with others, by near-monocultures, highly concentrated patches subject to frequent disturbance. The guilds of animals most immediately impacted by farming, therefore, would be graminivorous species and their predators, faced with extirpation through loss of habitat, or adaptation to a much closer association with people. Other impacts on woody-vegetation communities would be the intermittent but heavy impact caused by the removal of timber for construction purposes, and the less intensive but more frequent impact caused by removal of firewood.[79] Species of woodland and scrub habitats thus face a particular challenge: unlike graminivores, their habitat will be fragmented and reduced, not replaced in a highly concentrated patch. In fact, what agriculture in forested regions tends to do, apart from reducing the total forest cover, is to increase the forest-edge ecotone, albeit an ecotone subject to human disturbance at the edges of the human-modified environment. In much of temperate Eurasia and in the woodlands of eastern North America, one of the ongoing environmental stories of the Holocene has been loss of tree cover, fragmentation of woodland habitats, and the extirpation or adaptation of species reliant on that habitat.

Another impact was the construction of more or less permanent settlements. Although hunters and foragers need a place to call home, and some Paleolithic groups generated megasites such as Pavlov I (above), it is really only with agriculture that we see the widespread construction of residential settlements. Again, this constituted loss of local habitat patches, but the buildings themselves were a new habitat patch, presenting opportunities for any domiciliary species that could find a way to live within the spaces and interstices of the structure without attracting too much attention from its human neighbors.

The largest expressions of the built environment are the towns and cities in which most people now live. The origins of what could be called urban settlement go back to the prehistory of the Middle East and the city-states of Mesopotamia.[80] In turning to the Roman world for a case study, we are not choosing the oldest urban settlements nor necessarily the largest. However, the archaeology of the Roman Empire, especially the part of it that occupied what is now Western Europe, has been investigated in depth and in detail, and there is a very considerable body of evidence to investigate. Furthermore, this is not a world of discrete urban centers, political microcosms such as Ur or Great Zimbabwe, but a highly connected Imperium linked by road infrastructure and political hierarchy to a single center of power. It is easy to overstate the homogeneity of the Roman world, and to lose the local flavor of, say, North Africa, Gaul, and Britannia, but all of those regions were connected in practical as well as administrative ways, with the towns and cities representing the nodes on that network.[81] There was the potential, therefore, to acquire a commensal fauna in part by adaptation on the part of local species, and in part by deliberate or inadvertent importation of animals through trade and exchange networks.

Roman Britain makes a convenient and appropriate case study. The province of Britannia, though far from the Roman heartland, became quite densely populated, internally well-connected by a road system that can be traced even in the modern road network, and with numerous cities, smaller towns, and military bases of urban scale and density.[82] It is also a region that has seen frequent and thorough archaeological investigation, providing us with a substantial record to explore. The key questions are whether these Roman settlements developed an identifiable commensal fauna, and whether any such faunas were locally acquired, or imported as a part of the "Romanization" package.

Southern England had the first contacts with the Empire and became the most "Roman" part of Roman Britain. The commensal rodents that began their relationship with people in the Natufian of the Middle East found the Roman world to their liking. A few Iron Age records show that house mice had already reached Britain, so their presence in a number of Roman towns is hardly a surprise.[83] Rats, conversely, reached Britain during the Roman period, and are well-known from Roman London and from Dorchester.[84] Further north, rats are quite abundant in York from the second century onwards, and a number of other urban locations have yielded specimens.[85] The absence of records of rats and mice is complicated by the need for particular circumstances of deposition and recovery by sieving: we have to be careful to distinguish the absence of rats and mice from assemblages in which they might reasonably be expected—for example, those in which other rodents and small vertebrates are present. That said, the record from sieved samples from Roman Winchester is interesting, having yielded only a few records of house mice and no definite rats.[86] Absence of evidence is not evidence of absence, of course. However, Maltby reports small vertebrate remains from 218 samples from various sites around Winchester, with only a few ambiguous "large rodent" limb bones that could tentatively be rat, but might equally well be of water vole *Arvicola terrestris*. For comparison, rats were positively identified in 12 percent of samples from second- to third-century York,[87] at which frequency we would expect some twenty-six records of rats from Roman Winchester. The low frequency of house mice is similarly unexpected: the comparable record from York would lead us to expect in excess of one hundred records from the Winchester samples. Frequencies of the two rodents from London are more similar to those at York than at Winchester. Kevin Rielly reports frequencies of occurrence of rat bones in archaeological samples from a range of London sites that are quite close to my own figures for Roman York[88]—rats in 13 percent of first- to second-century samples, and in 20 percent of third- and fourth-century samples. Winchester begins to stand out, differing in this respect even from nearby Dorchester. Perhaps there was something different about Winchester, something that made this Roman town less attractive to commensal rodents. York and London were both ports, which may have made them more susceptible to the introduction of animals on incoming ships and carts. The apparent disappearance of rats from post-Roman York may be consistent with this ongoing introduction, as it may show that rat populations could only be sustained by continuous "topping up" from source populations on the continent, leading to local extirpation when the frequency of overseas contact declined. In that model, we would have to suppose that Winchester was less intensively integrated with cargo and shipping movements than were York or London. This is not inconceivable, but would require more evidential support than a dearth of commensal rodents.

Another species notable for its scarcity is fox. The species has been recorded from eight major settlement sites across Roman Britain (Colchester, Exeter, Leicester, London, Silchester, St Albans, Wroxeter, and York), but only in very small numbers, and it is altogether absent from a great many more sites.[89] For some reason, foxes failed to establish themselves in urban populations in Roman Britain, and the species was probably not a regular feature of the everyday landscape the way it is for so many urban Britons today. Similarly there are infrequent and mostly sparse records of pigeons, other than at Dorchester and Caerwent. In the latter case, the unusual abundance of bones attributable to *Columba livia* probably represents the husbandry of "doves" for food and manure.[90]

Apart from the rodents, the other commensal animals in Roman Britain seem to have been birds of the crow family. Specimens of raven *Corvus corax*, carrion crow *C. corone*, and jackdaw *C. monedula* are found regularly at Roman sites.[91] Although all three species are plausibly commensal, questions have been raised about at least some of the finds of raven. Maltby[92] points out that raven bones are quite often found in articulated groups, sometimes in special deposits other than occupation debris, suggesting that their deposition was something other than the discard of waste. A wider study of ravens and crows in Iron Age and Roman Britain has gone a little further, drawing attention to some continuity of the practice of deliberately burying ravens and sometimes crows in features associated with agricultural productivity,[93] from Iron Age settlements through into the Roman period. Perhaps these corvids did have some special significance. If they became a regular feature of everyday landscapes during the Iron Age, living as commensals around villages and hill forts, then their adoption by people into something symbolic or ritual seems quite feasible. Commenting on the ravens found during his early excavations at Silchester, Cyril Fox suggested that they may have been pets.[94] Serjeantson and Morris would rather have it that they were of some ritual or ceremonial significance, though the distinction may, in the end, be less significant than it seems. In either case, a species that was probably a familiar component of the commensal community became culturally integrated, either as tame companion animals or as symbols, living metaphors representing some concept, principle, or deity. I do not see these two roles as necessarily mutually exclusive: even in my own Western European culture, choosing a particular breed of dog (or no particular breed, or no dog) is a social signal, and the eventual deposition of the remains of that dog may be highly ritualized.[95] If ravens were often buried in special ways and places, this might indicate that ravens were an appropriate propitiatory offering to "the powers that be." However, it could just signify that a particular raven had crossed the line from being part of the commensal fauna to being a member of the household. Interestingly, a survey of animal bones from dozens of Romano-British temple sites across southern Britain found very few records of raven or crow.[96] Just two sites in the survey—Nettleton and Bath—yielded any number of bird bones other than domestic fowl. Regrettably, King does not comment on the identity of these bones, though even if they included corvids, the general dearth of corvids at temple sites would seem to leave open the question of their symbolic or companion status.

In all, the impression that one gains from Roman Britain is of successful populations of commensal rodents in at least some towns, the house mice having already successfully colonized, and the rats arriving as part of the "Romanization" of Britannia, with corvid birds making up the only other commensal group. Even if many of the raven records and maybe some of the crows are attributable to ceremonial deposition, that leaves some crows and most jackdaws, and it would be difficult to argue away all of the corvid records from Roman Britain as "ritual." Apart from those species, the archaeological record indicates that foxes made little use of urban spaces, and that feral cats were probably not a feature of Roman towns. Cat bones are certainly recovered from Roman sites, but only in quite modest numbers, at least in comparison with their frequency and abundance in the medieval period. Much the most commonly recovered small vertebrate from Roman towns in Britain is the common frog *Rana temporaria*. In all, this is almost an impoverished commensal fauna. In seeking an explanation, we might look to the nature of the Roman archaeological record itself. Without wishing to overstate the case that the Romans brought sanitation to the uncivilized Britons (and no writer of Irish descent would venture such an opinion!), it is the case that Roman archaeological deposits are often cleaner than those of later periods. There is more evidence of organized refuse disposal by drains and sewers,[97] and considerably less of the dumping, in every sense, that often characterizes even temporarily open spaces in medieval towns. Pottery and tile scatters indicate that refuse was carted away and dispersed onto fields in the surrounding area,[98] so it may simply be that most Roman towns offered less in the way of accessible food, at least for the larger commensal animals. Only the relatively small rats and mice, and perhaps frogs, stood to benefit, being able to utilize building cavities as living space, and being more able to access stored foodstuffs and refuse in covered sewers.

To sum up, it would be naive to assume that commensal associations between people and other vertebrate animals are a phenomenon of comparatively recent times. We are not surprised to see some animal species associating themselves with the spread of Roman towns, perhaps because that is an environmental context that is familiar to us today. However, people have been a significant force in changing environments and creating new habitat patches at least since the advent of agriculture 10,000 years ago, and perhaps arguably for many millennia before that. Crucially, the technological, cultural, or even cognitive capacity of the human population is not important, because commensal relationships may have been initiated largely by the other species. Would-be scavengers probably adapted to the feeding opportunities generated by Middle Paleolithic Neanderthals as readily as they would have done had those opportunities arisen through the activities of literate, modern people of the last few centuries. In that important sense, studying the archaeology of commensal animals is different from that of domestic animals, and the time depth of that study may be appreciably deeper.

4

Mesomammals

HAVING LOOKED AT COMMENSALISM AS A LIFE STRATEGY, AND HAVING CONSIDERED its possible time depth, we turn to some of the species involved. We need to consider what we know of the present-day ecology and ethology of those species: what is it about their behavior and way of living that has made them successful commensals? For some, their association with people may be quite recent in origin, an adaptation to the built and modified environment as we know it today, while for others, as we have seen, the historical and archaeological records show a much longer association with us. We may also consider whether some species adopt commensalism only locally or only temporarily, taking up the opportunity only when local circumstances make it possible or favorable. Terms such as "opportunistic" and "behaviorally flexible" are commonly applied to successful commensal and synanthropic animals, so we should beware of generalizing ecological or behavioral characteristics to all populations of a species at all times and in all places.

We begin with the mammals, termed *mesomammals* here to reflect the fact that all are of medium body size, too big to hide in crevices or under floorboards, but not big enough to be a physical challenge to people (usually). This size range could be summarized as too big to hide, but small enough to pick up. We start with possibly the most ambiguous species in this book, the cat.

CATS—MYSTERIOUSLY FAMILIAR

Cats are unusual among the species discussed in this volume in being conventionally regarded as "domestic" animals, or as companion animals or pets. However, cats occupy a range of places in our living space: for every carefully tended domestic pet there will be a less housebound cat that comes and goes *ad lib.*, and many feral cats that have long since given up any pretense of being human companions, yet that rely on our built environment and garbage for shelter and food. Not that this reliance is anything more than opportunistic. Cats that are deprived of their human support—for example, when people move away, or when cats are stranded on offshore islands—generally adapt remarkably well. Cats have even accompanied some expeditions to polar regions: the demise of Mrs. Chippy on Shackleton's *Endurance* expedition is a particularly sad tale.[1] The archaeological record seems to show that the fireside pet has its origins in an opportunistic commensal, one that we chose to adopt more closely into our lives. We think of cats as charming, enigmatic, and infuriating; studying their association with us may be all of those things. It may also tell us a lot about commensal animals as a whole, and about ourselves and our attitude to uninvited guests.

As a family, the Felidae includes some of the most highly specialized of terrestrial predators, superbly adapted animals, the larger species of which are a byword for ferocity in many modern cultures.[2] Alone of the thirty-seven extant species of felid, African populations of the wildcat *F. silvestris* Schreber have given rise to a form that we call "domesticated," a form that Linnaeus chose to call *Felis catus*. With that nomenclature, we run into a biological conundrum. Wildcats and domestic cats are formally regarded as separate species under those two names,[3] yet individuals of the two species clearly recognize one another as potential mates, and interbreed with enthusiasm to produce fertile offspring. On grounds of mate recognition, and because they are fully interfertile, it could be argued that the two forms of cat should be recognized as a single species, one species with two ecomorphs. However, if that were the formal position, it would be very difficult to define "wildcats" for conservation purposes, and European populations of wildcats, in particular, need active conservation precisely because interbreeding with feral cats has drastically reduced the numbers of "pure" wildcats.[4] The present nomenclature may therefore offend taxonomic purists, and is perhaps not strictly correct on a narrow definition of what constitutes separate species, but is certainly necessary in order to facilitate the conservation of wild, distinctly antanthropic cats. Even in matters of zoological taxonomy, cats manage to be mysterious and enigmatic.

Seen from an ecological perspective, cats are significant, biologically, for their complex trophic relationship with people and with other commensal species that form the food web in and around human habitation.[5] Cats generally constitute the top predator in that food web, making them a significant control on scavenging omnivores such as rodents and small birds that could otherwise become pest species from the human perspective. Unlike some other carnivores, cats appear to be adapted to a feeding regime of frequent small meals.[6] This reflects their solitary habits; there is no pack competition at a kill encouraging a cat to eat as much as possible. This behavior is seen in housecats, too, which prefer to have *ad lib.* access to food from which they take a modest amount (one mouse-worth?) at frequent intervals. Studies of feral cats on isolated islands[7] typically show that cats will focus on an optimal prey, such as rats inadvertently introduced by people, with little tendency to switch prey or to broaden the diet in response to variations in the density of the favored prey. As a result, feral cat populations in the absence of human-derived food may tend to fluctuate in cycle with those of their favored prey, in contrast to, for example, foxes (see below). This consistency of behavior should not be overstated, however. When a diverse range of small prey is available, cats may actually broaden their dietary range (contrary to the predictions of textbook optimal foraging models) in order to exploit the abundance of, for example, insect prey.[8]

We are familiar enough with the idea that cats are useful to people as a means of pest control, though there is an interesting chicken-and-egg question whether people are useful to cats as a means of attracting and aggregating the small prey on which cats live. In addition, cats also readily adapt to scavenging a wide variety of foods from human garbage, a tendency that may be more marked in higher latitudes such as northern Europe.[9] It is not clear whether this flexibility of diet is a commensal adaptation to living alongside people, or an exaptation from a trait present in early Holocene wildcat populations. It would be a mistake to assume that the cats that lived around the Middle East at the time when people were first settling in villages, farming grain, and attracting mice and sparrows, necessarily showed the same range of behaviors as the depleted and fragmented wildcat populations of today. Early Holocene cats may have been pragmatic in their behavior, perhaps more opportunistic in scavenging kills made by other predators. It is easy enough to see how that behavioral repertoire could readily adapt to scavenging garbage and accepting food handouts from people who found cats to be useful and appealing to have around.

A number of writers have made the case that cats were probably "self-domesticating," though none so clearly as Todd,[10] who argues that there was no credible process by which people could have deliberately initiated and managed the domestication of cats, and little reason for people to have done so. Thus, argues Todd, the impetus for cats to adopt a commensal habit must have come from cats exploiting the feeding and shelter opportunities offered by human settled communities. The pest-control benefits of cats may be overstated: field observations show that although cats may have a limiting effect on the growth of rat and mouse populations at lower densities, they are ineffective at reducing abundant rodent populations. Cats also tend to prey on voles in preference to mice when both are abundant.[11] In this model, cats began as opportunistic predators around human settlements, taking advantage of local concentrations of small mammal prey also attracted by the feeding opportunities.[12] As settlements grew, it became possible for cat populations to be supported largely by scavenging and some predation. A third stage was reached when scavenging was largely replaced by deliberate handouts of food from the human population, a point that may only have been attained relatively recently, and not by all housecat populations. From the human perspective, the transition from tolerated scavenger to deliberately fed "domestic" animal may be a significant step. For the cats, food handouts are just a more convenient and predictable form of scavenging. The predictability of food, both in time and space, is a recurring factor in the commensal adaptation of other species (e.g., see foxes, below).

Figure 29. Cats are not always what they seem. The slim, alert tabby is a thoroughly domestic "pet," while the plump, relaxed gray is one of the feral cats that live in the Minoan Palace at Knossos, Crete. (Source: author.)

From the human perspective, the attraction of having cats around may have extended to more than just pest control. As cats made the transition, at least in some places, from tolerated commensal to companion animal, other benefits may have become apparent. The role of companion animals in alleviating social stresses, with consequent benefits to measurable individual health parameters such as blood pressure and stress hormones, has been well documented, and cats, being furry, medium-sized, and facultatively self-reliant, are particularly well adapted to the companion role.[13] Cats may also have been symbolically significant for their familiar presence in the quotidian human landscape, and for their symbolically important role in moving between the constructed, managed *domus* and the "wild" world beyond. They are one of the most familiar of everyday species—in some cultures actively encouraged to share our homes, in others at least passively tolerated in feral populations. The frequent appearance of cats in statuary and other art, in early written sources, and in symbolic form as the manifestation of deities is an indication of their familiarity in past cultures throughout Eurasia and North Africa.[14] We can readily understand how people adapted to cats.

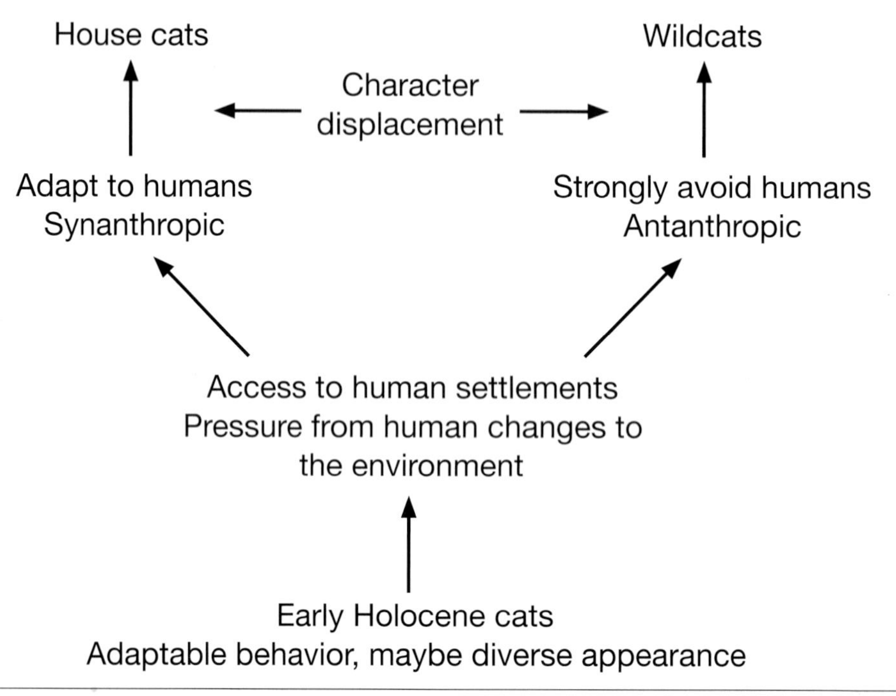

Figure 30. Association with people may have affected the behavior of the ancestors of commensal housecats and wildcats.

How have cats adapted to people? Given what we know of the biology and behavior of modern cats, we can offer some reasoned postulation. The process of coming together outlined above required relatively little behavioral adaptation on the part of the cat. Perhaps the most important was a reduction in intraspecies antagonism. Wildcats, like most felids, are essentially solitary animals, intolerant of the presence of other cats within their territory, whereas housecats will tolerate the proximity of other cats, particularly related individuals, and will live in startlingly high population densities when sufficient food is predictably available.[15]

Given that the defense of territory is largely in order to defend an exclusive hunting range, the availability of concentrated prey and scavenging opportunities around human settlements must have changed the selection pressures relating to this particular behavioral trait. In short, human settlements reduce the need for cats to defend an exclusive territory, and make a reduction in intraspecies antagonism positively beneficial. Again, observations of modern wildcat behavior in this and other regards need to be treated with some caution. In many parts of the world, wildcat populations have been living sympatrically with housecats for several millennia, and some behavioral character displacement is likely to have happened.[16] Studies of affiliative behavior in small felids kept in zoos seem to show that *Felis silvestris* is no more likely to show affiliative behavior towards humans than any other small felids, though there is some indication that African wildcats show greater affiliative behavior than other wildcat subspecies.[17] It is debatable how much studies such as this really tell us about the original coming together of people and cats. Studying the tendency of cats, pumas, civets, and so on, to relax their fear and avoidance of humans when in captivity tells us something about the specific setting in which the observations were carried out, but not necessarily much about the fundamental behavioral niche of the species concerned.

The model of cat "domestication" that emerges is of an uninvited dinner guest that became a tolerated lodger, then a member of the family. That is uncontroversial; the timing is more tricky. Phylogenetic studies of present-day wildcats and housecats show quite clearly that all modern housecat populations ultimately derive from Middle Eastern *Felis silvestris lybica.*[18] Less helpfully, estimated divergence times for the *lybica* and *catus* mitochondrial clades are around 131,000 years. This would place the domestication of cats around the beginning of the Last Interglacial, which seems improbable. More likely, the mitochondrial record is complicated, with repeated recruitment of maternal lineages from wild populations during a prolonged period of separation of the "wild" and "domestic" forms.[19] Although phylogenetic histories are often presented as "trees," evolutionary relationships within species or among closely related species are likely to have involved repeated introgressive hybridization, giving a structure more akin to a network than to a tree.[20] Thus, modern DNA evidence confirms the likely location of cat domestication as having been entirely within the range of Middle Eastern wildcats, though not the timing of that probably drawn-out process. The point is that this process could have happened at any time from the formation of the first human settlements—that is, at any time from the Natufian onwards. As noted in chapter 3, other opportunistic commensal species, such as mice (*Mus* spp) and sparrows (*Passer domesticus*), were evidently rapid colonizers of Natufian and Neolithic settlements in the Levant,[21] and these two familiar neighbors are discussed in detail later in this volume. There would seem to be clear grounds for expecting the predators of those small commensals, including cats, to respond to the consequent concentration of prey. We may suspect, therefore, that the association of cats with human settlements, and thus the origin of housecats, dates back to the beginning of the Holocene.

Convincing archaeological evidence of early domestication is hard to come by, even for species such as sheep and goats, for which control of breeding was probably established from the outset. For cats, there would have been little reason to control breeding, and little opportunity (it is difficult enough today without resort to surgery). Furthermore, the desirable characteristics of those early housecats were probably those of their wild relatives—predation and self-reliance—giving little reason to expect any morphological change consequent upon the closer association with people. Of the criteria commonly used to diagnose domestic status in zooarchaeological finds, morphological change is probably the least useful to apply to cats, at least in Africa and the Middle East. It is problematic enough to tell the two taxa apart in archaeological samples of recent date.[22] Two other criteria are also used: change of distribution beyond that of the presumed wild ancestor, and depositional context showing social incorporation different to that of the wild ancestor.[23]

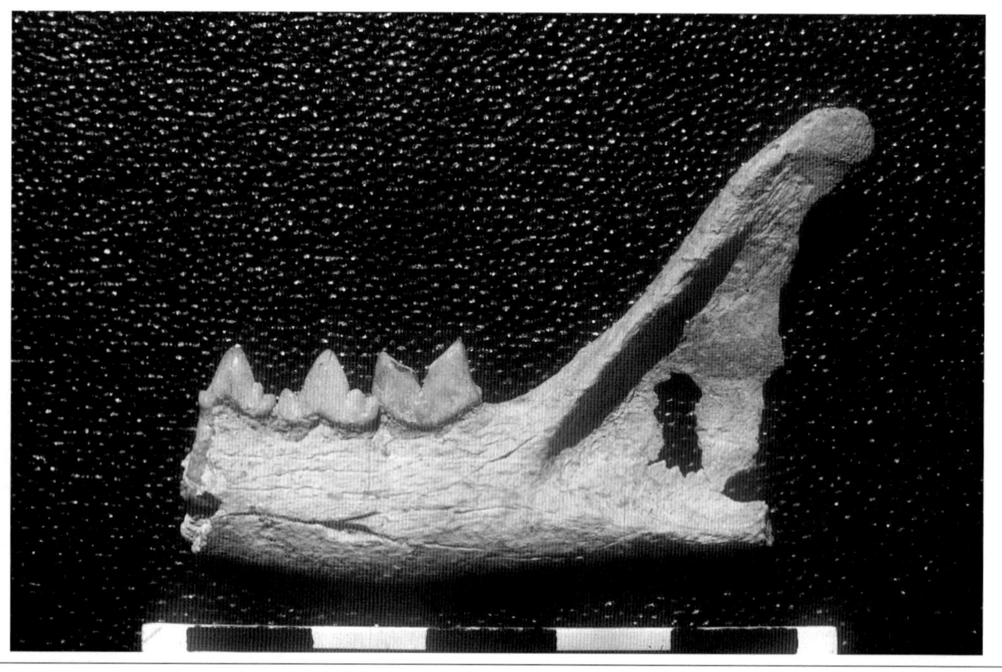

Figure 31. Neolithic cat from Khirokitia, Cyprus, showing that cats accompanied people to the island during prehistory. (Source: Simon Davis, with permission.)

Prehistoric translocation of cats has been convincingly demonstrated in the eastern Mediterranean. The island of Cyprus has been isolated from mainland Eurasia by 60–80km of water throughout the Holocene, and has no endemic felids. Accordingly, Simon Davis[24] was able to argue that a cat found at the Neolithic site of Khirokitia must have been transported to Cyprus by people. The same argument has more recently been made regarding a deliberately buried cat found at the Neolithic site of Shillourokambos (about 9500–9200 cal. BP), a site that has yielded other specimens of cat in Neolithic contexts and a feline figurine.[25] The status of these cats is somewhat debatable, and Jean-Denis Vigne and his colleagues make the point that the large size of the buried cat differs from the more evidently domestic cats found in later sites in Cyprus. However, the specimens from Khirokitia and Shillourokambos show that a close association between people and cats existed in the eastern Mediterranean early in the Holocene, and show that a sea crossing was no barrier to the distribution of cats. If the occurrence of a species beyond the range of its presumed wild ancestor is a criterion for recognizing domestication,[26] then it applies in the case of Neolithic Cyprus, and therefore, presumably, to the Levant mainland with which Cyprus had close cultural connections throughout the Neolithic. Vigne also comments that feline figurines have been found in Neolithic sites throughout the region, indicating some special status for cats, the social incorporation required as a criterion of domestication. Though not prime facie evidence that they were domesticated, such treatment at least indicates that cats were held in different regard than other carnivores.

The biggest difficulty that confronts an early Holocene date for the close association of people and cats is the prevailing account that has cats entering a domestic relationship first in Egypt—variously in the Middle or New Kingdom, depending on the source. Frederick Zeuner's influential account has been frequently reiterated, and he dates the first evidence to tomb illustrations of the 18th Dynasty.[27] In a nice example of the difficulties posed by this topic, Zeuner ascribes earlier zooarchaeological records, including those from Jericho and Harappa, to wildcats, but then expresses surprise that "the cat should have been domesticated at so late a period."[28] Malek takes a similar approach to early records of cats in Egypt—for example, describing the cat found at the feet of a Predynastic human burial from Mostagedda as "perhaps his pet,"[29] yet proposing that the first evidence for domestic cats comes from the much later 12th Dynasty tomb of Baket III at Beni Hasan.[30] In this tomb, a cat is shown facing an animal described in adjacent text as a rat, though in appearance it could equally well be a mongoose. Of course, a pet need not be a domestic animal per se, and that point could be made about the cats from Neolithic Cyprus. However, the Beni Hasan illustration simply shows a cat (identified in text as female *mit*) in a frieze, other registers of which show numerous people and several baboons. Are the baboons domestic animals? If not, is the cat possibly "just" a pet, like the Mostagedda cat? Zeuner illustrates an 18th Dynasty scene that shows a cat, a duck, and a monkey all apparently romping beneath a chair. The cat is illustrated as a wild-type striped tabby, but Zeuner nonetheless infers domestic status for the cat from the lively domestic scene, though apparently not for the monkey. The point in all of this is that there has been a tendency to interpret cats shown in "domestic" settings as housecats, while excluding cats found in close association with people but at an appreciably earlier date than the earliest "domestic" illustrations. Zeuner and Serpell[31] both support an Egyptian origin for housecats by citing etymological evidence that the terms that have entered English as "cat" and "pussy"

Figure 32. Although much debated, this cat on a papyrus stem (Egyptian Middle Kingdom, Beni Hasan Tomb3) has been argued to show a domestic cat. (Source: Plate 5 from F. Ll. Griffith, ed., *Beni Hasan*, part 4, *Zoological and Other Details* (London: Egyptian Exploration Society, 1900).)

are of North African or Near Eastern origin, via Latin. A detailed critique of this etymology is beyond both the remit of this book and the competence of its author (not to mention the patience of its readers). Suffice to say that the survival of words possibly derived from Ancient Egyptian forms does not gainsay the much earlier existence of housecats known by names that are long since lost, in languages of which we have no record. The earliest recorded words that we have for the hippopotamus are also Egyptian, yet the creatures certainly existed at a much earlier date, and presumably their contemporaneous humans called them something.

The Out-Of-Egypt model for cat domestication then distributes cats via the Nile Delta to the Greek world, thence to Rome, and into Europe by way of Roman conquest. Given the ambiguities that attend recognition of housecats in the textual, pictorial, and zooarchaeological records of Egypt and the Middle East, at least the OOE model is amenable to objective testing. If this model is to be accepted, there should be no convincing pre-Roman records of housecats in Europe beyond the Mediterranean basin. The key word there is "convincing," as pre-Roman cats in northern Europe will fall into the same trap of acceptability as Predynastic Egyptian cats have done. Diagnosis must proceed by careful consideration of depositional context, which may be difficult for old published records, and by testing the biometric evidence for the presence of two morphologies. On this second point, at least, the challenge in northern Europe is to distinguish housecats derived from African wildcats from endemic European wildcats. In fact, it appears that housecats reached even the British Isles in pre-Roman times, with several records from Iron Age settlement sites in mainland Britain,[32] and possible records from the Orkney Islands, beyond the northern limit of Roman influence.[33]

It is a reasonable assumption that North American populations of housecats derive from animals taken across the Atlantic by European settlers from the seventeenth century onwards. That said, there is a fascinating sliver of evidence to suggest that earlier European colonists may have left their traces in modern American cats. Analyses of the genetics of cat populations all across the North Atlantic region show that cats from the northeastern United States have more affinity with cats from Ireland, western and northern Scotland, and western Norway than with the rest of North America or with mainland Britain.[34] Although this study, now rather dated, wants support from present-generation techniques, it continues to raise the wonderful possibility that the Norse settlers of Vinland brought cats with them, and that those cats persisted more successfully than did their human companions.

Cats have persisted on offshore islands, in the absence of a prime facie commensal niche, with such success that it is almost more notable when they have failed to establish. John Long records cats on almost two hundred islands around the world,[35] and on many of these they have become a significant threat to often rare and endemic species of ground-nesting birds.[36] Although Long, writing in the 1970s and 1980s, noted only a few cases where feral cats have been successfully eradicated, mostly around New Zealand, eradication programs have been put in place for many more islands in recent years. Nogales and colleagues report success on at least forty-eight islands, though only ten of these are larger than 10 km^2. In the subantarctic regions, cats established a feral population on Macquarie Island, midway between New Zealand and the Antarctic continent. A program of eradication was put in place from 1985 to 2001, in order to reduce predation pressure on native birds in this World Heritage site. The unintended (but surely not unforeseeable) consequence has been a boom in numbers of rabbits, causing landscape-scale changes in vegetation and thus in soil erosion.[37] At a similar latitude, cats have retained a paw-hold on the icecaps and crinkly coastline of Kerguelen Island, where they keep the rabbit numbers under some degree of control. Somewhat unexpectedly, cats failed to thrive on the Hebridean islands of St Kilda. After the beleaguered human population was removed in 1930, twelve cats were released onto the main island. By 1931, eleven of them were dead, so the twelfth was shot—rather unfairly, one feels. No doubt a number of factors contributed to this failed colonization, though the marked seasonality of prey may have been one significant factor. St Kilda famously abounds in seabirds from April to August, but lacks small prey through the colder months of the year. The famous Soay sheep were presumably beyond the capacity of even the most desperate and ambitious cat. Cats attuned to preying on birds for part of the year may simply not have been able to make the switch to scavenging sheep carcasses and tide-line corpses in the winter. Commensal cats have the advantage that our homogenized, modified environment tends to show low seasonal variation in the provision of food, whether scavenged (garbage is present year-round) or hunted (so are mice).

Figure 33. Mrs. Chippy, the ship's cat on Shackleton's 1914–17 Endurance expedition to Antarctica, accompanied by Perce Blackborow, youngest of the crew.

A full account of cats and people would require a book on its own, and would go well beyond the remit that this volume has set for itself. The apparently ambiguous archaeological record may, in fact, be quite easily resolved if we accept that cats had a lengthy prehistory living in and around our settlements as commensal animals prior to whatever reciprocal moves on the part of their human neighbors changed the status of at least some cat populations from wild-but-commensal to domestic. The underlying thesis of this book is that the conventional division of "wild" and "domestic" animals does not reflect the range and subtlety of relationships between ourselves and other species, and cats illustrate this point rather well. Maybe Zeuner was right after all, and "domestication" in the full sense of people having control over the animals' breeding really did not happen until New Kingdom Egypt, having been preceded by more than five thousand years of cats living alongside us, expediently developing the affiliative and socialized behavior that is so unlike the behavior of wildcats, yet without becoming a controlled part of our socioeconomic systems.

Several millennia later, our attitude to cats is becoming ambiguous. On the one hand, cats as pets are an industry worth millions of dollars, pounds, euros, and yen annually. On the other, cats are seen as a problem, even a menace. Margaret Slater sees this most clearly in attitudes to feral cats, and the increasing tendency to regard feral (i.e., unmanaged and unsocialized) cats as part of "nature," and therefore a suitable subject for control and management, rather than as animals that share our living space and with which we coexist.[38] For some, the only appropriate control is eradication, while others advocate TNR (trap, neuter, release) as means of population control. The perceived need to eradicate feral cats in vulnerable island environments has already been touched upon. Closer to home, control of housecats is now advocated as a means of conservation of small birds.[39] Both in Europe and in North America it is argued that housecats and feral cats have an adverse effect on populations of small songbirds in and around settlements, and some quite disturbing numbers are adduced in support. Dauphiné and Cooper, for example, estimate that cats kill one billion birds per annum in the United States, and one study in Dunedin, New Zealand, estimated that the annual "harvest" of six bird species by cats was close to the estimated total population for those species.[40] Other studies have drawn more nuanced conclusions, usually along the familiar lines of "more research is needed." For example, a year-long survey of cats at ten sites in Bristol, UK, showed that almost 60 percent of them returned no bird prey throughout the year (compared with 21 percent in the Dunedin study), and that the birds that were caught were, on average, in worse physical condition than the bird populations at large.[41] That would seem to show that cats act as compensatory rather than additive mortality, though an appreciable cull of birds was noted in the Bristol study. It is to be hoped that further research on this issue will not polarize into pro- and anti-cat lobbies, despite the signs of that happening: advocacy makes for poor research. Taking a *longue durée* view of cats and people, we can see a 9,000-year trajectory from tolerated commensal to desirable domesticate to conservation menace.

Figure 34. Commensal interconnections between cats and bees can be seen in my garden. The bumblebees benefit from the presence of cats, which prey on the mice that might otherwise prey on the bees' nests. The food I put out for birds, however, is also consumed by the mice. The bee is feeding on chives *Allium schoenoprasum* grown for human consumption. *Cui bono*? (Source: author.)

To conclude, an admission of vested interest: I share my home with two cats. I also put out food for wild birds, which probably more than compensates for the cats' occasional predatory activities. In fact, both cats tend to catch mice rather than birds, and by reducing the numbers of field mice around the garden and adjacent field, they help to conserve the solitary bees and bumblebees that are such a feature of the garden in summer. Painting commensal animals as wholly good or bad is just not that simple.

CANIDS AND RACCOONS

A similar tension between amiable neighbor and villain arises with the commensal canids. Like the cats, the dog family, Canidae, has developed a wide range of relationships with people. Most obviously, dogs have become a domesticated animal, apparently derived from the Eurasian grey wolf. There is some discrepancy between the genetic and zooarchaeological evidence on this point. Specimens of large canids attributable to dog, as against wolf, are dated to the European Upper Paleolithic (see chapter 3), yet genetic studies of modern dog and wolf populations seem to indicate a much older divergence between the two, possibly as long as 100,000 years ago.[42] That is a little startling, as it would imply that wolf and dog began to diverge more or less as modern *Homo sapiens* was moving from Africa into the Eurasian home of grey wolves. If that 100,000 year antiquity stands up to further research, then a close association between people and wolves began almost as soon as the two species came into regular contact. Alternatively, this may be the same issue that we encountered with cats: a long period during which there was introgression to the "dog" gene pool from "wolf" matrilines, producing a reticulate, intraspecific evolution that is not easily resolved into a branching phylogeny with a clear-cut "divergence date." Fascinating though this topic is, dogs are domestic animals, and so beyond the remit of this volume. However, we should not rule out the possibility that there has been a long history of mutual facilitation, even a degree of commensalism, which influenced the behavior and evolution of at least some wolf populations sufficiently to have contributed to that early apparent divergence date.

Despite our species having its origins in sub-Saharan Africa, it is not clear which, if any, native African species adopted commensalism. The African wildcat appears to be the ancestor of domestic cats, so in one sense that species is an example, though genetic studies show that domestic cats derive from wildcat populations in the Middle East, not in sub-Saharan Africa. An interesting candidate species is the black-backed jackal *Canis mesomelas*. Jackal bones are commonly identified in assemblages from Late Stone Age and later sites in Africa, and it would be easy enough to regard them as just another prey animal. For example, at the Abbots Cave site, in the Seacow River valley of the South African Cape, the frequent jackal remains were discussed in terms of the ethnographic evidence for the use of jackal skins.[43] However, even if the jackals were skinned, the presence of appreciable numbers of their bones indicates that the animals were procured close to the cave—i.e., that they were quite common close to the occupied cave site. There may be some parallel here with the occurrence of foxes at Natufian sites (chapter 3). Furthermore, there is intriguing evidence from some recent DNA work. Horsburgh extracted DNA from the bones of animals believed on anatomical and contextual grounds to be early specimens of domestic dogs from southern Africa. The DNA analysis showed all of them to be black-backed jackals.[44] The immediate significance of this study was that it showed how readily bones of two similar canid species may be confused, and that DNA analysis may be used to distinguish them. Going a step further, the canids were identified as dogs in part because of the contexts in which they were found, so the DNA analysis leaves us with jackals ("wild" animals) in a domestic context. It would be premature to add the black-backed jackal to the list of successful commensal animals, but the evidence should certainly prompt a reconsideration of the relationship between people and this widespread animal. Jackals are diverse feeders, actively predatory as well as scavenging and adapting to a wide range of dietary items. In a predator guild that included the lion, leopard, cheetah, and the larger golden jackal *C. aureus*, black-backed jackals may sometimes have gained benefit from associating with locations regularly occupied by hunting peoples. The species certainly adapts well to the presence of people: it has continued to be abundant and widespread despite greatly increased settlement of southern and eastern Africa.[45]

Figure 35. An urban coyote making good use of a public park. Is this becoming a typical coyote encounter? (Source: Janet Kessler, coyoteyipps.com, with permission.)

Other canids are clearly commensal, having moved into our settlements for food and shelter while remaining outside the realm of domesticity. In Europe, we are now familiar with urban red foxes, commensal populations that add metropolitan flair to the fox's reputation for

Figure 36. An urban raccoon leaving its favorite diner. (Source: Worth's World, with permission.)

being cunning (as in Janáček's opera) or simply fantastic (as in Roald Dahl's children's story). In North America, several canid species have adapted to living synanthropically around towns and cities, notably the coyote *Canis latrans*[46]—though whether any of these populations can be described as commensal, as well as synanthropic, requires further discussion. The same applies to a number of other medium-sized mammals that have found a way of life in the urban world, such as raccoons *Procyon lotor*, the Virginia opossum *Didelphis virginiana*, and the striped skunk *Mephitis mephitis.*[47] But it is foxes that have most clearly adopted a commensal way of life, and therefore where we begin.

Red foxes extend over much of the northern hemisphere, and have been introduced with varying degrees of success into Australia.[48] Foxes share many of the characteristics of successful commensal mammals. They are omnivorous: rural foxes take a wide range of vertebrate and invertebrate prey, and urban foxes have quickly developed a taste for pizza and noodles. A survey of the stomach contents of 402 foxes shot in the city of Zurich, Switzerland, between January 1996 and March 1998 showed that over half of the food in a typical stomach was directly human-derived.[49] Incidentally, the mere fact that over four hundred foxes could be culled out of this urban population over a 27-month period without putting a significant dent in the population shows how abundant and resilient fox populations can be in cities. Even in rural areas, foxes will take advantage of human activities that concentrate food. In Poland, in a region of extensive farming with low human population density, foxes will target chickens as a food source even though rodents and other prey are abundant.[50] Their opportunism as predators is seen particularly in regions of Spain where rabbits are abundant. Comparisons of fox diet with rabbit abundance show a very simple relationship: when rabbits are abundant, foxes target rabbits, but when rabbits are scarce, the foxes simply eat other things, with little indication that rabbit abundance affects fox abundance.[51] Foxes breed rapidly, with a typical clutch size of four to six young, and first breeding at about ten months old.[52] Mortality is quite high. Most free-living fox populations show a typical life span of only a few years, while captive animals may live to ten years plus.

Over much of its range, the red fox has close relatives as competitors. Its role as a pest is therefore somewhat constrained, as is its capacity to adopt and exploit a commensal niche. Within northwestern Europe in general, fox is the only small endemic canid.[53] Since the extinction of the grey wolf over much of its former range, fox lacks a larger-bodied close relative as a limiting factor on its distribution, numbers, and behavior. Interactions between fox and badger *Meles meles* are complex: both are omnivorous and they are of similar body size—though an adult badger will outweigh a fox, badgers being more heavily built. The two species seem to avoid direct contact quite successfully, possibly because both use scent marking to label their territories and routeways. It is easy enough for a human, even with our feeble olfactory sense, to detect when we are in fox or badger territory.[54] Foxes could thus have a relatively unconstrained niche in the British Isles and adjacent Continental Europe but for a long tradition, particularly in England, of fox hunting.

Fox hunting is a subject that arouses strong passions. A deceptively simple issue—the acceptability or otherwise of using a pack of hounds to track and kill foxes—it unpacks into questions of social class, the power of landowners over the landless, and the extent to which "nature" is completely excluded from modern farming. Fox hunting has powerful social resonance within twenty-first-century England. Some of its consequences may be cryptic, yet far-reaching once understood. As the farming and land-owning aristocracy took over and remodeled the landscapes of lowland England in the later eighteenth and nineteenth centuries, they constructed an idealized rural landscape within which agriculture could satisfactorily take place, but which was also directed towards other country pursuits, including fox hunting. Even today, the English countryside can be read to show that fox hunting was important enough to influence the construction of the rural landscape.[55]

A full treatment of this fascinating topic lies beyond the remit of this book. However, the long history of fox hunting in England has given the red fox a distinctive symbolic role, a "personality," which has influenced attitudes towards foxes as they have moved into our towns and cities over the last few decades. On the one hand, foxes are villainous, breaking into hen houses

Figure 37. Traditional English fox hunting depicted by Henry Thomas Alken (1785–1851).

and killing fowl in numbers well beyond their immediate food needs. This pattern of killing, though often exaggerated, is nonetheless a disagreeable aspect of genuine fox behavior.[56] It is probably a sensible response to an abundance of prey: kill many, eat some, then remove and cache carcasses for future consumption. The fact that it is often described as the fox killing "for pleasure" or "out of sheer spite" says more about our propensity for anthropomorphism than it does about foxes. Conversely, foxes are "talked up," perhaps in order to make it seem that the pack of hounds and a dozen or so chaps on horseback are pursuing a quarry worthy of them. Foxes are characterized as cunning and intelligent, resourceful, inventive. They are depicted as an essential countryside character—one which, nonetheless, is there to kill and to be killed. Most perversely, foxes are not infrequently depicted wearing the distinctive costume adopted by (human) fox hunters. The brewers of Old Speckled Hen, a bottled beer popular in the North of England, take that association one step further by using dressed-up foxes in the advertising of their product. The point is that foxes had a very well-known image when they started moving into towns and cities in England. That image has worked both for and against them, their role as cunning and resourceful vermin running up against deeply embedded differences between urban and rural England.

Exactly when foxes moved into English towns is a moot point. The 1930s spread of the suburbs, with their large, leafy gardens, may well have offered foxes an excellent opportunity, though it is only in the last few decades that the "urban fox" seems to have become a legitimate subject for study. Detailed surveys in the 1980s showed that foxes were well established, with a distinct preference for better-off suburban neighborhoods rather than city centers or local authority housing.[57] The population of urban foxes in the UK is usually put at around 33,000, though that is out of a total fox population estimated at 258,000.[58] By and large, town dwellers in England like to see foxes in their neighborhood. A poll of 4,000 households held in 2010 showed that 65.7 percent liked to see foxes around town, 25.8 percent had no view either way, and only 8.5 percent actively disliked urban foxes.[59] However, that was before an extraordinary incident in Hackney, East London, in which nine-month-old baby twins were apparently attacked and injured by a fox while asleep in their own home.[60] The fox seems to have entered the house during the late evening through open patio doors, which is not in itself unusual on a warm summer evening in London. The parents were alerted by noises from an upstairs room, in which they found the two babies in their cot, badly bitten about the face and arms. There was a fox in the room, which fled the scene unharmed. Two things are remarkable about this case. The first is that it is absolutely without authentic precedent.[61] Foxes may become habituated to people, suppressing their usual flight response, but direct aggression towards people is very rare, unheard of in situations where the human has not made the first aggressive move. When confronted, foxes back down, and they are readily seen off, for example, by a domestic cat.[62] The second notable aspect is the polarization of views that this case has generated in the media. Advocates of country "sports" have rushed to point out that "townsfolk" simply don't understand what vicious and dangerous vermin foxes really are, and blog sites have fairly buzzed with anecdotal tales of the "foxes ate my pet rabbit" variety.[63] Against this, fox biologists and soft-hearted nature lovers have combined to point out the uniqueness of such an incident, and to ask that urban foxes in general should not be misjudged because of the actions of one exceptionally aberrant individual.[64] Only a very few have quietly pointed out that dogs attack children far more often than foxes do; that Hackney abounds in dogs, not all of which are well-trained family pets; and that the "fox in the room" may be a cover story to protect a neighbor's dog. One should not forget, of course, that two small children were badly hurt in this incident, though both have recovered well following hospital treatment. Whether the babies were attacked by a dog for which the fox stands patsy, or by a fox exhibiting extraordinarily aberrant behavior, the case has opened up a public debate about urban foxes—a debate that has shown just how deeply nineteenth-century attitudes to rural vermin persist in twenty-first-century urbanized England.

Other European countries have urban fox populations, and surveys of attitudes give results quite similar to those seen in England. A thorough questionnaire survey of one relatively prosperous suburb of Munich, Germany, for example, prompted responses to quite a range of questions.[65] The great majority of respondents were pleased to see foxes in their gardens, and 91 percent agreed with the statement "Foxes have a right to live." Only a small minority (8 percent) expressed concern that foxes might injure family pets, though it is an odd statistic that a larger minority (15 percent) were concerned that foxes might injure people.[66] In fact, the larger source of concern was the health hazard posed by tapeworms carried by the foxes. Foxes carry *Echinococcus multilocularis*, and urban populations may have a high rate of infection.[67] This parasite can be acquired by people who come into close contact with fox feces, potentially causing a significant lung disease. The majority of Munich suburbanites want to see foxes in their gardens, but want the foxes to be wormed. This desire for management is consistent with 78 percent agreeing that "Because foxes have no natural enemies, people must re-establish the ecological equilibrium."[68] Bearing in mind the mortality data from many urban-fox surveys, one might observe that foxes are heavily preyed upon by road traffic, so people already are a highly intrusive factor in fox population ecology.

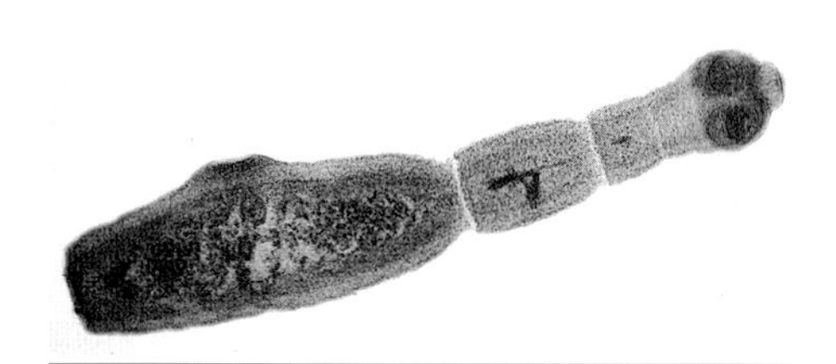

Figure 38. *Echinococcus multilocularis*, a parasitic worm carried by foxes and a potential health hazard to their human neighbors.

The archaeological record gives us a nice example of the changeable relationship of foxes and people. Foxes are absent today from the Orkney Islands, off the northeast coast of Scotland, yet they appear in the archaeological record of those islands for a limited period of time. Sites dated to the Iron Age and succeeding "Pictish" period, roughly from the last few centuries B.C. to the sixth and seventh centuries A.D., yield relatively abundant remains of foxes,[69] while earlier and later sites do not. Absence of evidence is always tricky, but the abundance of foxes in the Iron Age in contrast with its dearth before and after is quite striking. But why? One possibility is that people had little to do with it: foxes may have succeeded in colonizing Orkney from the adjacent mainland of Scotland—a glance at the map shows only 13km of sea separating the two. Therein lies the danger of maps. Those few miles of sea are the Pentland Firth, a seaway notorious for being cold and stormy, and a place of ferocious tides and currents. Natural colonization across the Pentland seems highly unlikely, not least because foxes show little evidence of colonizing successfully across seaways of their own accord. Foxes are absent from a number of off-shore islands around Scotland that would seem to be at least as accessible as Orkney. And if "natural" colonization were feasible, why the restriction to a few centuries, and not a series of colonization events to maintain a more or less continuous population? A more plausible explanation, then, is that foxes were deliberately introduced into Orkney. Their appearance in the archaeological record coincides with the period of *brochs*, great stone tower-houses that, whatever their other functions, certainly displayed the power of their occupants. And the extirpation of foxes coincides with the colonization of those islands by the Viking settlers whose names and language are such a feature of the islands today. Eva Fairnell proposes that foxes were introduced and managed for their fur, a distinctive contrast to the grays and dull browns otherwise available from seals and domestic livestock.[70] Some fox specimens from Orkney, from the second and third centuries A.D., show cut marks consistent with their having been butchered. The extirpation of foxes may then show a marked change in how they were regarded by the people of Orkney. It seems reasonable to ask whether foxes were established as a commensal animal, hanging around settlements, scavenging off the refuse heaps that dot Iron Age sites in Orkney, and preying on sea birds and their eggs. That a few were taken from time to time for their skins would not be at all inconsistent with this way of life. Their extirpation in Viking times may have occurred less through a cessation of management than through deliberate hunting of a species now perceived as a pest. Over a few centuries in late prehistory, foxes in Orkney may have undergone a full transition from desirable introduction to tolerated commensal to extirpated vermin, a microcosm of their trajectory in mainland Britain as a whole.

An unexpected result from the archaeological record is a real paucity of records of foxes from medieval towns. The image that we have of urban life in medieval England is far from salubrious. In part this derives from a few written sources that describe food waste, excrement, and butcher's debris littering the streets, in part from the laws and ordinances that sought to remedy the problem, and in part from the archaeological record itself, which largely consists of garbage dumped in backyard and brown-field areas.[71] The abundance of rodents and urban corvids is consistent with this abundance of free food, but where are the foxes? Given the success of urban foxes in recent decades, it seems obvious that they would have found a desirable home around the squalid streets of medieval England. The paucity of fox records must be seen relative to those of other urban commensals such as rats, corvids, and kites, which occur far more frequently in the animal bone record. One possibility is that foxes have only developed the commensal habit in certain populations quite recently. Looking at the modern evidence, that would seem possible, as the spread of foxes into towns in England is largely a phenomenon of the last fifty years.[72] However, the archaeological record shows foxes to have been associating with people at some times and in some places for millennia (chapter 3), suggesting that their opportunistic and adaptable tendencies are nothing new. A possible explanation may lie in the abundant records of cats from those same medieval towns. In a few cases, these cats have been deliberately buried, and we are probably considering the remains of a much-loved companion. The great majority of cat records, however, are disarticulated bones mixed into the general urban refuse, more consistent with untended feral populations of cats. If most medieval towns had a substantial population of feral cats, that may have inhibited colonization of the towns by foxes: competitive exclusion. That interpretation would predict that foxes would be excluded from towns unless feral-cat populations were reduced long-term. It may be circumstantial evidence, but the increase in urban foxes in England over the last few decades matches the decline in substantial feral ("stray") cat populations.[73] In reducing the numbers of stray cats living around our towns and cities, we may have given the foxes just the opportunity that they lacked in medieval times.

RACCOONS AND OTHERS

One of the fascinating aspects of animal adaptation to anthropic environments is the enthusiasm with which medium-sized carnivores (mesocarnivores) have taken up the challenge. Urban areas in most parts of the world have their characteristic commensal mesocarnivores, quite apart from cats and foxes. Probably the best known of these species is the raccoon *Procyon lotor*. Raccoons are native to North America, though they have now been introduced into parts of Europe and Asia. They are largely nocturnal, opportunistic omnivores, adapted to a wide range of habitats, but probably originally animals of wet woodlands.[74] Their life span and mortality is similar to that of foxes: raccoons can live well into their teens, but in the wild, few live beyond about five years. They have one litter per year, usually of three to seven young, and reach sexual maturity at about one year old for females, two years old for males. In many respects, raccoons are absolutely typical mesocarnivores, with the particular adaptation that their forepaws are highly dexterous, enabling raccoons to manipulate small objects with great skill. The epithet *lotor* refers to their habit of "washing" food items.

Studies of raccoon behavior in urban and suburban settings in North America and in those parts of Europe where they are now becoming quite common consistently show adaptations to commensal living that are also seen in other mesocarnivores. Studies in Illinois have shown that raccoons in urban locations have smaller home ranges than their rural conspecifics, show less seasonal variation of home-range size, and live at higher population densities.[75] The reasons are not hard to fathom: the distribution and abundance of human-derived food (e.g., garbage, stored products, accessible pet food) strongly modify raccoons' normal foraging behavior, leading to foraging over much smaller areas than in rural settings, focused on aggregations of food.[76] Very much the same thing has been observed in raccoons in Europe, including the observation that urban raccoons show a high degree of overlap in individual home ranges.[77] Perhaps this is another reflection of the reduced intraspecific antagonism noted in feral cats? In passing, a study of raccoon behavior in western Poland also noted the versatility and sheer cheek of one raccoon living "wild" in a forested area, which on several occasions adopted a disused sea-eagle nest 35m up a tree as a daytime resting place.[78] Other studies in North America have shown that urban raccoons demonstrate higher site fidelity than rural ones (i.e., they are significantly more likely to be recaptured or observed at the same location), and that the ratio of juveniles to adult females tends to be higher at urban locations, perhaps reflecting larger litters.[79]

Some of the observations made for urban raccoons are replicated in other mesocarnivores when living in close association with people in heavily modified environments. The same high population density and significantly reduced home range has been noted in badgers *Meles meles* living in the seaside town of Brighton, in southern England.[80] In this case, it is debatable whether the badgers could be properly described as commensal, as the feeding advantage that they gain is from "natural" prey of invertebrates and small vertebrates concentrated in rich garden habitats, rather than from directly human-derived food. Badgers are a recent addition to the urban fauna in the UK; whether they will adapt their behavior to include raiding bins and stealing food put out for pets remains to be seen. In parts of mainland Europe, the stone marten *Martes foina* shows the same adaptation. Studies in southern Luxembourg have shown just the same higher population density, reduced home range, and greater home-range overlap as in urban raccoons.[81] In stone martens, however, the effects are appreciably less marked, leading the authors of this study to point out the dangers of overgeneralizing across mesocarnivores, and the role that phylogenetic traits may play in moderating the response to environmental opportunities.

Commensal behavior in mesocarnivores is not restricted to urban environments. In nature reserves and game parks that have a high throughput of human visitors, dense concentrations of garbage at picnic sites or around refuse bins may create exactly the right conditions to facilitate aggregated feeding by any opportunistic species that cares to seize the moment. (Older readers will at this point recall Yogi Bear . . .). At the Queen Elizabeth National Park in Uganda, for example, refuse bins and other, mostly small, garbage dumps around staff residences are regularly raided by banded mongoose *Mungos mungo*.[82] The bins provide a healthy mix of meat, eggs, and rice, and also serve to attract the invertebrate animals that would normally feature in the diet of this species. It is perhaps no surprise, then, that refuse-fed adults were on average heavier and in better physical condition than adults from the same population that did not regularly refuse-feed. Female refuse-feeders were found to carry more fetuses on average than non-refuse-feeding females, but did not on average have larger surviving litters. An interesting reflection of the social challenges represented by aggregated feeding is that young male refuse-feeders showed a higher mortality than any other class of young. In all, access to garbage certainly had a range of effects on the health and reproduction of the mongoose population, but it would be overgeneralizing to suppose that the effects were necessarily always beneficial.

African mongoose populations have plenty of "natural" predators to trim their populations and to affect their behavior. Reduction of predator pressure is one possible benefit of adopting a synanthropic, commensal lifestyle. In some parts of North America, even in quite urban locations, a potential predator exists in the form of the coyote *Canis latrans*. Stanley Gehrt has documented the increasing presence of coyotes in urban neighborhoods, noting that urban coyotes tend to be more nocturnal than their rural conspecifics, and to occupy smaller home ranges.[83] It is a reasonable supposition that the presence of coyotes would affect the density, home range, and behavior of raccoons and other mesocarnivores through interference competition if not outright predation. Remarkably, systematic observation and experimentation seems to show that the presence or absence of coyotes makes no difference to urban raccoons[84] or to skunks.[85] The ecological and behavioral consequences of adapting to aggregations of human-derived food would seem to override any effects from the presence of a larger competitor or predator.

Mesocarnivores are an interesting and familiar group of animals, perhaps because so many of them have adapted to living within and around our settlements. One region of the world in which they may be more significant than we currently realize is Southeast Asia. This fascinating cluster of land masses from continents to tiny, remote islands has a complex and rich biogeography, which archaeological work is beginning to extend into the past. Certain species have clearly colonized even remote oceanic islands by close association with people, and the travels of *Rattus exulans* are discussed in a later chapter. The mesocarnivore niche in this region is occupied by species such as the diprotodontid marsupial cuscus *Phalanger orientalis* and the palm civets *Paradoxurus hermaphroditus* and *Paguma larvata*. Their present-day distribution suggests that these species were transported to islands by earlier human populations. Cuscus appear in the archaeological record of New Ireland by 19,000 years ago, and are highly unlikely to have made that sea-crossing from New Guinea without human intervention.[86] Cuscus seem also to have been transported into the Solomon Islands by 9,000 years ago.[87] As both translocations would have been undertaken by hunter-gatherer peoples, presumably the intention was to establish offshore populations of a useful prey animal. Cuscus are largely frugivorous and herbivorous, and nocturnal in habits.[88] The more omnivorous *Paguma larvata* is kept in households around island Southeast Asia today as a ratter. It is an altogether likely candidate to have been a commensal mammal that attached itself to human settlements, because that is where the rats were most abundant, and then adopted and transported about as a useful neighbor. Although palm civets have not acquired quite the niche in Southeast Asia that cats occupy in much of mainland Eurasia, it is quite plausible that an initial stage of commensalism followed by social incorporation occurred with both species.

One other group of mammals that has been closely associated with people in Southeast Asia is the monkeys, a group whose relationship with people has been distinctive and close, both in evolutionary terms and as animals that have fascinated people in many parts of the world.

MONKEYS

As a group, monkeys capture the imagination. They are fellow primates with familiar faces—intelligent, curious, and noisy. Our attachment to them ranges from adopting monkeys as companion animals, to recruiting them as coworkers in fruit-picking or street music, condemning them as pests of fruit crops, utilizing their similar biology in medical research, and trying to conserve the rarest and most vulnerable of them in the wild. For their part, a number of monkey species have adapted to living commensally with human populations around the more tropical parts of the world. There is a chronological perspective, too. As an African primate, humans and their ancestors have shared space and resources with Old World monkeys (Cercopithecidae) throughout the 6 million years of evolution of our clade, and paleoanthropologists are intrigued by the possibility of commensal or facilitational relationships between ancestral hominins and their monkey contemporaries, such as the extinct baboon *Theropithecus oswaldii*.[89] Today, monkeys have been forced into ever-closer contact with people as tropical forest habitats have been split up and reduced in area. For some, this has been disastrous: of twenty primate species listed as "critically endangered" by the International Union for Conservation of Nature (IUCN) in 2006, five are Old World monkeys and three are apes.[90] For others, people have provided living space and feeding opportunities; indeed, the interaction between people and monkeys is sometimes so close and complex that one wonders whether they recognize us as close relatives as much as we recognize them.[91]

Figure 39. A Langur monkey demonstrating house-breaking skills, India. (Source: Philip Piper, with permission.)

Three groups of monkeys have adapted particularly well to living around people: the macaques (*Macaca* spp.), the baboons (*Papio* spp.), and the vervets (*Chlorocebus* spp.), of which the first two groups have shown the greatest facility to adopt commensalism. Macaques are a particularly interesting group of monkeys, with a wide geographical range that includes the least tropical environments in which monkeys are to be found today. With their wide distribution and ecological versatility, macaques are perhaps a better analogue for the human family than any of the great apes. Japanese macaques *Macaca fuscata* are famous for their use of warm springs to alleviate winter cold, and for apparently deliberately making snowballs.[92]

In more tropical regions of Asia, macaque populations live within towns and at rurally located tourist sites. In these situations, the monkeys commonly make the transition from exploiting handouts and stored food to aggressively begging. Once again, a successful commensal animal becomes a little too successful, and comes to be regarded as a "problem." A recent survey in Shimla municipality, India, investigated the ecology and behavior of two species of monkeys in a region densely populated by people.[93] Both rhesus monkeys *Macaca mulatta* and Hanuman langurs *Semnopithecus entellus* were well habituated to living close to people, but the survey showed quite clearly that only the rhesus monkeys had adopted a commensal way of life. Distribution of the two species was subtly different. Rhesus monkeys were recorded in 22 out of 25 municipal wards in Shimla during the survey period in 2008, but langurs in only 11 of 25. Commensal feeding was directly observed in nearly half of the forty-nine rhesus monkey groups, but in none of the fourteen langur groups, despite the survey evidence that both species were accustomed to, and comfortable with, the regular presence of people. It is no surprise that direct harassment of monkeys by people was observed frequently for the rhesus monkeys and only intermittently for the langurs. What the Shimla study nicely shows is the importance of species biology and ethology. Give populations of two monkey species the same challenges and opportunities, and they will not necessarily respond in the same way.

However, the Shimla case may be reflecting langur behavior only when sympatric with macaques. Tom Waite and colleagues report commensal behavior in Hanuman langurs in the Indian city of Jodphur.[94] Jodphur lies at the western edge of the distribution of this species within India, and marginal populations are often the most susceptible to local extinction through, for example, inimical climate events. Between 1999 and 2001, a major El Niño–Southern Oscillation (ENSO) event caused greatly reduced rainfall across India. Hanuman langurs in the Kumbhalgarh Wildlife Sanctuary, Rajasthan, were badly hit and reduced in number by about 50 percent. The commensal langurs of Jodphur, meanwhile, showed negligible change in numbers over the same period, having been buffered against the consequences of the ENSO event by handouts of food. Waite points out that the Jodphur langurs are by no means eking out a living on urban detritus, but instead have a quite benign living environment.[95] Whatever other motives may be driving human feeding of langurs, religious devotion plays a major part. In Hindu symbolism, monkeys are sacred to the deity Hanuman, and the black facial skin of the langurs strongly associates them with stories of Hanuman being burned in an act of heroism. For some people, therefore, feeding the langurs is a devotional act. Given the antiquity of Hinduism, it is beyond the reach of the historical record to show whether the belief came first and led to the protected and subsidized status of the langurs, or whether some langur populations had already adopted a commensal association with people and were co-opted into the emerging belief system.

Religious significance of monkeys in general is reflected in attitudes within India to the ubiquitous gangs of commensal rhesus monkeys. Rhesus macaques are famously assertive. In a twelve-month period on the Guahati University campus in Assam, India, 27 cases were reported of rhesus monkeys biting people, and a further 49 cases of "aggressive threat."[96] The threats were predominantly directed towards women, not men, though the published study does not compare those frequencies with the ratio of men to women in the campus human population. Rhesus monkeys were recorded snatching food directly from people's hands and entering houses in order to access food, often causing damage in the process. Reaction to the monkey problem was nonetheless moderate. Asked about remedial measures, most residents preferred that the monkeys should be translocated (presumably without discussion of how and to where!), with a minority supporting neutering of males or other contraception procedures, and only a very few agreeing that shooting the monkeys would be best. The similarity of these survey results to European surveys of attitudes to urban foxes is quite interesting. Whether or not belief in the sanctity of monkeys continues to be widely held, it may underpin a more secular presumption against killing monkeys, even when they are an abundant, aggressive pest. Phyllis Lee and Nancy Priston point out that this divine status under Hinduism is matched in Chinese and Japanese traditional cultures by a respect for monkeys as a manifestation of devious cunning, leading to a form of love-hate relationship between people and monkeys across South and East Asia.[97] In many places within the region, this conflict is managed by encouraging the feeding of monkeys within a temple precinct, while they are shot as pests in neighboring fields.[98] There is another echo here of the English attitude to foxes, also respected for their cunning, yet hunted as destructive vermin, and another example of spatial boundaries being drawn to define where "nature" is and is not acceptable.

Baboons are widespread across tropical Africa, and their distribution intersects with human settlements and activities of all kinds. The capacity of baboons to become habituated to people is (in)famous: visitors to drive-through wildlife parks are warned not to stop—not for fear of lions or other potential predators, but because baboons will mob cars, causing appreciable damage.[99] Roadside baboons are a familiar sight in many parts of Africa, and it is no surprise that they have adapted to become adept and bold crop raiders. Field observations of habituated, commensal baboons tend to yield conflicting results—for example, disagreeing over the periodicity and regularity of crop raids.[100] This suggests that baboons are highly adaptable, varying their "technique" according to the opportunities and circumstances. A detailed study of one group in Nigeria has shown just how well adapted they are. Crop raids were successful on 69 percent of observed occasions. Farmers took an average of twenty-three minutes to react to baboon raids,

Figure 40. Roadside baboon *Papio cynocephalus* in Tanzania, where baboons have learned that passing cars often slow down and deliver food handouts. (Source: author.)

and baboons are capable of stuffing quantities of food into cheek pouches. This allows them to enter fields, forage quickly and quietly (some studies indicate a reduced level of vocalization among foraging groups in cultivated fields in comparison with foraging in "the wild"), fill their cheek pouches and hands, and head away well inside the likely human response time. In the Nigerian study, 52 percent of crop raids were terminated by the baboons with no human intervention, with less than one-third being successfully interrupted by farmers chasing the baboons away.[101] Although there is little comment on this point in the research papers concerned, these observations would seem to indicate that baboons have a sense of time, some internal clock that enables them to estimate when the limits of a safe foraging period are approaching. In short, baboons are successfully commensal, and there seems to be little prospect of people reducing that success.

Attitudes towards commensal animals in general can be highly culture-specific, but towards monkeys they are deeply complex. Monkeys resemble us, not only facially but behaviorally, and that has no doubt been a factor in their incorporation into myth and religion, which in turn feeds into secular attitudes and reactions. There are ecological overlaps, too. Where agricultural settlements move into monkey habitat, those settlements are often growing crops such as carbohydrate-rich seeds (maize, millet) and fruits such as mangoes that are central to monkey diets. In the wild, the availability of such foods will fluctuate seasonally, while cultivated crops are likely to be managed to provide a year-round sequence of produce.[102] The benefits for monkeys are evident. Socioeconomic factors will influence attitudes to crop raiding. In subsistence farming at its most basic, some degree of loss to crop raiders may be tolerated and accommodated by growing a little more than is necessary. However, in a cash economy, any crop surplus beyond the needs of the family or village has value, so there is no excess to be tolerantly shared with the monkeys. We forget this important point when we project present-day attitudes to commensal animals back into the past. Landscape changes may also influence degrees of tolerance. Deforestation in Sri Lanka has had a marked effect on people's willingness to tolerate commensal monkeys. Where substantial areas of continuous forest remain close to villages, the human inhabitants remain quite tolerant of the monkeys; but in villages around which forest has largely been removed, that tolerance is significantly reduced.[103] In part, this difference in tolerance probably reflects a situation in which the monkeys are perceived to have "moved in" permanently, rather than visiting regularly from the forest—though one wonders whether there is also an underlying sense that a largely deforested landscape has been made by and for people, whereas settlements on the forest edge are in a more negotiable position.

Most parts of the world have acquired their own commensal mesomammals. Some of those animals are essentially carnivores (foxes, coyotes), while others are mainly herbivores or frugivores (monkeys). What they have in common is the capacity to adapt their diet to take advantage of a wide variety of foodstuffs—whatever human crops and garbage offer. Some larger mammals may become commensal locally and opportunistically, such as bears, even polar bears. However, those circumstances are uncommon and locally contingent, and people's tolerance of them is understandably low. Large body size implies large territory and a large sheltered space in which to sleep and breed. A combination of all of those constraints will act against large mammals regularly adopting a commensal habit. Coyotes are probably at the upper end of the adaptable size range, which may in part account for the uneasy nature of the relationship between coyotes and people where they co-occur (further discussed in chapter 7). Notwithstanding the alleged baby attack described above, foxes are

unlikely to pose a direct threat to a person, whereas a coyote that has become h
people is a different prospect. We have seen the part that culture-specific belie
moderating the tolerance shown towards particular species, and the subtle factors that
fine-tune that tolerance. From cats to coyotes to langurs, quite an assortment of mesomammals have found their place within and around our settlements, and have been doing so, and doing well, for millennia.

5

Rats, Mice, and Other Rodents

EXPLAIN "COMMENSAL ANIMALS" TO MANY PEOPLE AND THEIR FIRST RESPONSE IS "Oh, like rats and mice?" These species are all too familiar to urbanized Western societies, but perhaps their familiarity leads us to neglect and underrate them. The association of rodents with people goes back over the millennia, extends all over the world, and includes species other than the familiar house mouse and common rat. Rats and mice are biologically highly successful, and their prevalence today is something of a monument to their versatility, and to the cozy niche that they constructed thousands of years ago.

Rats and mice are rodents of the zoological family Muridae, a family that is originally Eurasian but is now worldwide in distribution—in part because its many species have successfully adapted to a wide range of environments from the wet tropics to the arid Middle East, and in part because a few murid species have hitched a ride with people, using our homes and food stores as places to raise an abundance of offspring, and co-opting our means of transport to follow us all over the globe.[1] Although a great many rodents have been deliberately or accidentally transported by people, it is the murids that have been most successful in this regard, and four species in particular deserve close attention: the house mouse (*Mus musculus*, *M. domesticus*); the ship rat, black rat, or roof rat (*Rattus rattus*); the common rat, brown rat, or Norway rat (*R. norvegicus*); and the kiore, Maori rat, or Pacific rat (*R. exulans*). A couple of other taxa will be mentioned in passing: the spiny mouse *Acomys cahirinus*, and the Australian swamp rat *Rattus lutreolus*. The African giant rat, or pouched rat (*Cricetomys gambianus*) is neither a murid nor a commensal within the terms of this book, though its increasing popularity as a pet and its valuable role in land-mine detection almost earn it a place. People's attitudes to rats and mice vary culturally and individually, ranging from acute, almost phobic revulsion, to the affection that many people in Europe and North America feel for pet common rats. The murids have become iconic in Western cultures, from Mickey Mouse to the altogether more streetwise and sardonic Roland Rat.[2] This chapter will review what we know of the long history of murids and men, beginning with house mice, moving on to rats, then reviewing the current situation for the group as a whole.

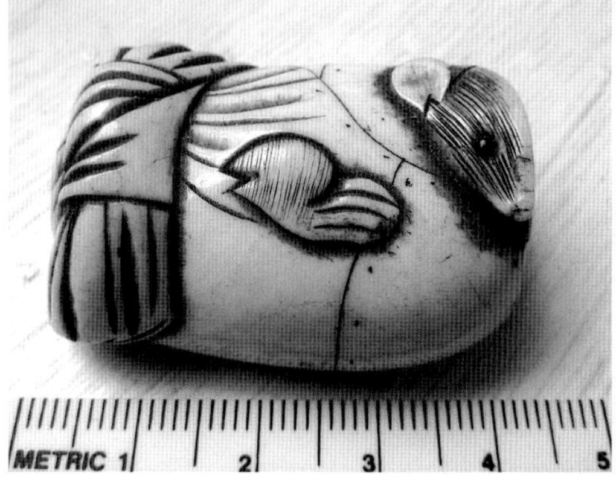

Figure 41. Cultural attitudes toward rodents are depicted in *netsuke*. The example left shows two playful mice, and right, a rat bundled up for sleep. (Source: Sonia O'Connor, with permission.)

HOUSE MICE: FROM THE FERTILE CRESCENT TO ANTARCTICA

First a word regarding nomenclature: the animals that we call house mice have been subject to a complex history of taxonomic revision.[3] Linnaeus started the ball rolling by naming the species *Mus musculus* (the little mouse). That name held until 1943, when Schwarz and Schwarz[4] tackled the diversity of house mice around Europe and southwest Asia and proposed splitting them into a series of species. In particular, they argued that a clear distinction could be drawn within Europe between an eastern species and a western species. As Linnaeus had named *M. musculus* on the basis of a specimen from an eastern population, the western species needed a name of its own. The earliest published candidate, and therefore the one to take priority, was *M. domesticus*, based on specimens from Dublin described by Rutty in 1772. Some authors regard the eastern and western forms as subspecies, hence *M. musculus domesticus* and *M. m. musculus*. Two other *M. musculus* subspecies should be noted. *M. musculus castaneus* occurs throughout Southeast Asia, where it is closely associated with people's dwellings, and *M. musculus bactrianus* is described from Iran to northern India—though of uncertain validity, as it is known from relatively few specimens and seems to be genetically remarkably diverse. For our present purposes, we can call them all house mice, and deal with their past and present on those terms.

So successful have house mice been in spreading around the globe that it is quite difficult to decide where they originally called home. Genetic studies place their center of divergence in Southwest Asia: indeed, the diversity of modern *M. m. bactrianus* populations may indicate the region from which other house-mouse stocks have dispersed. As we have seen with foxes and perhaps cats, archaeological records indicate that our association with mice began in the Middle East at the end of the last climatic cold stage, some 11,000 years ago.[5] People were beginning to settle in small villages as sedentary foragers who made good use of the wide range of seed- and nut-bearing plants to be found in the region. This economic model gradually underwent a transition as cereals and pulses were brought into active cultivation, to be followed at least a millennium later by domestic livestock. Although house mice are quite versatile feeders, the provision of fields and stores of grain and lentils proved to be highly attractive. By the time the archaeological record shows clear evidence that crop cultivation had become the economic mainstay, the bones of house mice become a common find on settlement sites.[6] The speed with which house mice adapted to take advantage of the new niche is quite remarkable. The major Neolithic settlement of Çatal Hüyük, in Anatolia, developed into a substantial accretion of mud-brick buildings packed side by side, apparently entered from the contiguous flat roofs (see fig. 27). In gaps between the structures, garbage accumulated—today excavated with care as important archaeological deposits, but originally a magnet for scavenging mice, whose remains are abundant at the site.

The practice of agriculture spread throughout the Mediterranean basin and into western Europe from its origins in the Middle East, taking domestic livestock and crops into new lands. It would be reasonable to expect that the mice traveled with this expansion, yet the evidence suggests that, at least in the Mediterranean, the dispersal of farming did not necessarily mean dispersal of house mice.[7] Archaeological specimens and modern genetics show mice entering Europe by two pathways. The "northern route" is the less well studied, taking mice through the western Black Sea region and along the Danube into eastern Europe—hence the distribution of *M. musculus* populations today. The "Southern Route" took mice through the Mediterranean basin to France and Spain, and hence into the westerly parts of Continental Europe, to become the *M. domesticus* populations that subsequently spread throughout the world. Thomas Cucchi and his colleagues argue for a three-stage spread of house mice by the southern route. An initially rapid and geographically limited dispersal moved mice throughout the eastern Mediterranean region, as outlined above. In the second stage, there is little evidence for successful colonization beyond this region from the seventh to the first millennia B.C. Despite some previous claims for a Bronze Age spread of mice, very few reliably dated specimens are known. An important exception is a solitary house mouse mandible from the Uluburun shipwreck of southern Turkey and dated to about 1300 B.C.[8] In the third stage, roughly from 1000 to 500 B.C., mice spread rapidly throughout the western Mediterranean and into northwestern Europe. In the Iberian Peninsula, the archaeological evidence shows house mice and house sparrows (*Passer domesticus*) moving northwards from the Mediterranean coast in association with Phoenician influence and trade.[9]

Why this late dispersal? What inhibited the spread of house mice into farming villages throughout the region at a much earlier date? Three possibilities can be conjectured, each of which has some relevance to wider questions of how commensal populations come to be spread around the planet. First, there is the question of transport. In order for viable new populations to become established beyond the original "wild" range, sufficient mice had to be moved, perhaps more than once, from a source population to a new location. It may simply be the case that there was too little maritime traffic through the Mediterranean in the Neolithic and Bronze Age to "seed" new locations with sufficient mice. Second, there is the question of niche. The Middle East was notable for the formation of villages and towns such as Abu Hureyra and Çatal Hüyük during this period of pre- and protohistory—the rest of the Mediterranean less so. Maybe mice were moved around, but simply lacked the particular constructed environment that their commensal adaptation required. Third, there may have been competitors. Neolithic and Bronze Age settlements in western and central Europe commonly yield bones of wood mouse (*Apodemus* spp.), even from samples taken within dwellings. Although wood mice are not thought of as particularly commensal today, the less constructed, more organic settlements of this region and period may have offered them an opportunity to exploit human settlements more than they do today. Having said that, wood mice show little hesitation in moving into the author's home during cold winters. In Neolithic western Europe, commensal populations of wood mice could have been regularly "topped up" from the wild, whereas house mice would have been isolated, and dependent on their founder population and its descendants. So, any combination of transport, niche, and competitors may have combined to keep house mice confined to the eastern Mediterranean. House mice clearly are susceptible to exclusion by competitors. In the Mediterranean Levant today, house mice share the region with a relative, *Mus macedonicus*. The two species occupy contrasting niches: house mice are restricted to dwellings and their immediate surroundings, while *M. macedonicus* lives in scrub and uncultivated land. However, further west in the Jordan Valley, where *M. macedonicus* is absent, house mice form permanent, "feral" populations away from settlements.[10] The commensal adaptation in house mice allows successful colonization of a region from which the species would otherwise be excluded by a competitor.

The last outpost in northwestern Europe for an itinerant rodent is the British Isles, far removed from the Middle Eastern homeland, climatically challenging, and relatively late to develop the built environment of towns and cities. The earliest records in Britain date from the last few centuries B.C., from hill forts such as Danebury and the major settlement at Gussage All Saints.[11] These records are a good reminder that Iron Age Britain was not some cultural backwater populated by woad-bedecked barbarians, but an outward-looking island that had trade and exchange with Continental Europe sufficient in volume and frequency to acquire unintended passengers. There are records of house mice just across the North Sea in previous centuries, and Iron Age sites in southern Britain are particularly characterized by pits and timber structures apparently intended for the storage of grain. The niche was available, a source population was nearby, and the mice just needed sufficient transport to set them in place. Competition with native wood mice may have been an issue, as records of *Apodemus* species are common enough in Iron Age sites in Britain. It may be that only the largest, most "constructed" settlements were sufficiently amenable to house mice and inimical to wood mice to hold viable populations. This is a conjecture that can be tested as a greater number of secure archaeological records becomes available.

If house mice were only patchily successful in Iron Age Britain, the Roman occupation must have presented many favorable opportunities for dispersal and increase. House mice are found at many of the Roman sites where bone preservation and recovery makes it possible for small rodent remains to have been recovered. Not all archaeological excavation makes sufficient use of sieving (screening) for us to be confident that records of absence are not merely absence of evidence, but chapter 3 shows that the evidence for house mice in the larger towns of Roman Britain is quite variable, indicating substantial populations in some towns and few in others. Presumably the degree and frequency of cargo movements between Britain and the Continent increased, at least periodically, facilitating the topping up of established mouse populations by new immigrants. With the effective collapse of urban living in the post-Roman centuries (fifth to eighth centuries A.D.), house mice must have faced something of a challenge. The archaeological record for this period in Britain is not substantial, but a few sites have shown that mice adapted and survived. In York, for example, house mice continue to be found in post-Roman deposits and on into the medieval period. At Caerleon, in South Wales, house mice were quite numerous among the prey of barn owls deposited on the floor of an abandoned Roman frigidarium, alongside wood mice and various shrew and vole species.[12] It would appear that some house-mouse populations, at least, found a modus vivendi that enabled them to maintain viable populations during the supposed "Dark Ages." The resurgence of towns from the eighth century onwards reestablished the niche, and house mice have been a feature of towns throughout northern Europe ever since. Genetic analysis of present-day house mice around the British Isles shows clear evidence of two waves of colonization contributing to the modern populations.[13] One is limited to the northern and western margins of the archipelago and includes haplotypes also found in Norway. These mice, it is proposed, originated in Norway and were moved around the Atlantic fringe by Viking traders and settlers in the ninth and tenth centuries A.D. The second group, typical of most of mainland Britain, has haplotypes more typical of the European mainland, particularly Germany, and probably represents the original Iron Age dispersal of mice into Britain. Interestingly, one British haplotype (BritIs13) has close affinities with North American house mice, supporting the historical evidence (below) that Britain was the main source for house-mouse populations in the United States and Canada. Using house-mouse genetics as a proxy record of human dispersal has been particularly effective in Sweden. House-mouse genetics in this part of Scandinavia intriguingly shows that they have *musculus*-type nuclear genes, but *domesticus*-type mitochondrial DNA (i.e., derived only from the maternal line). That combination is seen along the narrow zone from Jutland to the Balkans where the European ranges of the two forms overlap, yet in Swedish mice this "hybrid" genotype occurs much further to the east, well into what should be *M. musculus* territory. What this probably shows is that a quite small number of *M. domesticus* mice hitched a ride from North Germany into Sweden, probably as farming spread across the southern Baltic around 4,000 years ago.[14] The prehistory of commensal behavior in house mice is written in their genes.

The arrival of house mice in North America is surprisingly poorly documented. John Long, usually a reliable source on the dispersal of mammals around the world, attributes the first introduction to Stephen Harriman Long's 1819–20 expedition, specifically into Iowa.[15] Although Major Long traveled extensively along the Platte and Missouri rivers, he cannot have been the original source of European house mice in North America, even if he was instrumental in moving them from the East Coast right across the Midwest. Once again, genetics may be the most useful source of information. House mice on the East Coast share a distinctive genetic polymorphism with mouse populations in Britain and western France.[16] The same polymorphism is not seen elsewhere in Europe, so its presence in mouse populations in Florida indicates that these sunshine-seeking mice originate from British or French populations, not from Spanish mice. Based on estimated rates of evolution and divergence in mouse genomes, house mice arrived in North America in the second half of the eighteenth century, in good time to join Long's expeditions into the interior. Whether they arrived before or after 1776 remains unknown. Similar dates probably suffice for the arrival of house mice in Australia. They may well have accompanied the First Fleet in 1788, and will have accompanied the majority of European vessels ever since. Some Australian populations have successfully established themselves in a feral state,[17] independent of the commensal niche. In that sense, the species has been more successful in Australia than it has been in Britain, a situation not unfamiliar among emigré Brits in Australia. It is an interesting reflection on human traffic around Australasia that house mice in Australia are predominantly of the *M. domesticus* form, consistent with introduction from western Europe, whereas those typical of New Guinea are predominantly of the *M. musculus castaneus* type that is widespread in Southeast Asia.

Throughout the eighteenth to twentieth centuries, house mice accompanied people around the globe, traveling with cargoes, disembarking as the cargoes did so, and finding a modus vivendi wherever people built houses and stored food. They have colonized the Atlantic region from Iceland in the north to South Georgia in the far south. Much the same can be said of the Pacific, with house mice having been recorded in the Aleutian archipelago and, as an apparently quite vigorous population, on Macquarie Island, halfway between New Zealand and Antarctica. These pioneering mice probably arrived with sealers and whalers in the late nineteenth or early twentieth centuries, and clearly prospered. Subsequent well-intentioned introductions to Macquarie were of cats, to control the mice, and rabbits, to provide fresh meat for visiting mariners. The cats found the native bird fauna to be more to their liking and promptly had a devastating effect on sea birds, which presumably had never seen a mammalian predator before, let alone a number of hungry ship's cats on shore leave. The decision was taken in 1985 to eradicate the cats for the sake of the endemic birds, and by 2000 they were gone. Meanwhile, rats had arrived and were multiplying enthusiastically, replacing cats as the major predator of young sea birds on the nest, and the rabbits had done what rabbits do best and were destroying the fragile and distinctive Macquarie flora. Removing the cats to save the endemic birds had also removed the only predator of the rabbits, the rats, and the mice. Macquarie Island is quite a story.[18]

So, house mice have utilized their close relationship with people to colonize much of the land surface of the Earth, in a wide range of environments. What of their biology: do house mice have particular attributes that have made them so successful? House mice are famously prolific breeders. Females typically have four to eight young per litter, and can produce five to ten litters per year. Fecundity is closely linked to population density; in very dense populations, breeding may cease altogether. Commensal populations of house mice breed year-round, while feral populations tend not to breed during the winter months.[19] Life span obviously depends on predation pressure; house mice attain several years of age in artificial conditions, but this is a poor indication of the life expectancy of a typical mouse.[20] House mice are surprisingly pragmatic, adapting their social behavior according to the availability of space and food. This adaptability can be shown in modern house-mouse populations, and was probably an important factor in the adoption of a commensal niche by some mouse populations, as noncommensal populations typically show a higher incidence of agonistic behavior.[21] This same adaptation, being more tolerant of the proximity of other individuals, has also been proposed as an important step in adaptation to commensalism in cats.[22] It would seem that a successful commensal mammal must not only tolerate our species but also its own.

Our tolerance of mice may be another matter. They cause damage and fouling in stored foodstuffs and other organic materials, and to have resident mice in one's home is generally regarded as both a practical problem and a social black mark. Apart from their propensity to sample stored vegetables, house mice are quite adept at eating their way through modern packaging of most kinds. I recall unwisely leaving a bag of shopping on the kitchen floor in a mouse-infested house in the Orkney Islands, and returning to find that a tunnel had been gnawed (or consumed) right through the center of a packet of biscuits. In this particular house, the mice had made themselves at home by burrowing into a horsehair-stuffed chaise longue, proving that Orkney house mice have a certain *je ne sais quoi*.

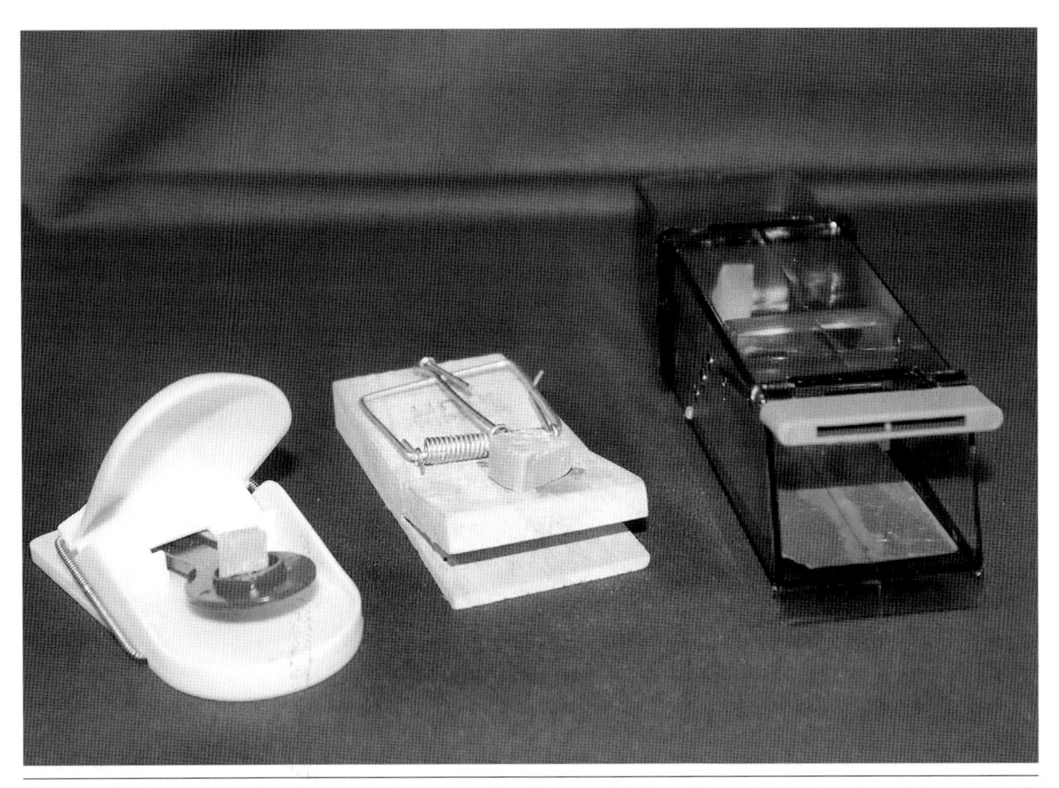

Figure 42. The Promax trap does the same as the conventional trap next to it, but in modern materials, and the Longworth live-trap, widely used in field studies of small mammals. (Source: author.)

It would be remiss to discuss the domiciliary habits of house mice without at least some mention of mousetraps. In a neat example of cultural coevolution, the adaptation of mice to homes has been matched by the production of ingenious means of trapping or killing the animals. Somewhat remarkably, mousetrap technology has even been used as a case study in technological evolution, as a counterargument to the Intelligent Design argument of irreducible complexity.[23] Although ceramic artifacts interpreted as live-traps for mice have been reported back to 2500 B.C.[24] there is surprisingly little archaeological evidence of mousetraps. A range of everyday artifacts could have been put to that purpose, so it is quite possible that mousetraps have a deeper antiquity and wider distribution than would seem to be the case. An early treatise on mouse traps was published by Mascall in 1590,[25] in which some thirty-four different traps are described, nine of them specifically for mice. Most rely on some form of baited trigger, which sets off a wire noose or drops a heavy weight on the unfortunate mouse. Mascall's "mill" is a fiendish device of freely rotating vanes that are baited and positioned such that a mouse seeking to feed from the vanes is likely to be precipitated over the edge of the table and into a pot of water, in which the mouse is drowned. William Fitzwater's entertaining and detailed survey of trapping[26] includes a number of unusual mousetraps, not least the Kness Ketch-All trap, a spring-wound gizmo that live-traps a mouse, throws it into an adjacent storage compartment, then resets itself for the next victim. To conclude the subject of trapping mice, the eminent zoologist Martin Hinton had quite firm views. He was of the opinion that mice are best attracted by some "delicacy rare in the locality," so traps set in a cheese shop should be baited with fish, and vice versa.[27] Eminence notwithstanding, Hinton was reputed to have an eccentric sense of humor, and his advice on trapping mice may or may not have had serious intent.

To be fair to house mice, though they may eat our food and infest our wainscoting, they have been remarkably helpful to humanity in their medical role. Although we may think of the medical role of mice in terms of testing therapies and modeling physiological processes, there is a remarkable amount of literature on the use of mice as medicine. The symbolic and therapeutic place of mice in Ancient Egypt is nicely summarized in an old and charming paper by Warren Dawson.[28] One aspect of the image of the Nile as a giver of life was the way that mice arose spontaneously from the Nile mud after each inundation. The source for this belief as authentically Egyptian, it should be pointed out, is Pliny's *Natural History*, which is sometimes more credulous than credible. In the terms used in chapter 2, Pliny is anecdote rather than science. However, the notion that the mud of the life-giving river could give rise to animal life just as it did to lush vegetation is plausibly Egyptian. Reisner's 1901 excavations at Nag-ed-Dêr (Nag' ed-Deir), in Upper Egypt, yielded a number of desiccated bodies from burials dated to the Predynastic period. The bodies were examined by the eminent Grafton Elliot-Smith, whose autopsies on Egyptian mummies and desiccated bodies did so much to instigate a new aspect of Egyptology, as well as depriving future generations of scholars of intact study material on which to work. Elliot Smith noted remains of mice in the alimentary canals of several children and concluded, in discussion with his eminent contemporary Fritz Netolitzky, that this was of medical significance.[29] The grounds for doing so are mentions in several medical papyri of the use of mice as medicines or as poultices, sometimes on their own and sometimes in combination with other animals. The Hearst papyrus, for example, mentions putting "cooked mouse in fat until it is rotten" in the preparation of a powerful medicine. Other sources recommend feeding cooked mice to children in order to stop them dribbling or to prevent bed-wetting. Pliny apparently drew on these Egyptian traditions and a range of other sources in listing a great many therapeutic uses for mice, and subsequent would-be authorities have drawn heavily on Pliny. Unsurprisingly, English sources into the late seventeenth century, including the famous Thomas Culpeper of *Herball* fame,[30] were still recommending the use of mice (skinned and cooked, naturally) for a range of disorders.

The Enlightened eighteenth century saw the decline of mice as medicine, at least in mainstream literature. In the early 1800s, we have the first records of house mice being captive-bred as laboratory animals, and by 1829, Coladon had been experimenting with cross-breeding different color strains in an effort to understand inheritance.[31] The first inbred strains of laboratory mice were developed at Harvard in 1909. Among the many strains of house mice still maintained for laboratory use, some date back to these early years of lab mice. These "old" inbred strains are notable for having only one haplotype of mitochondrial DNA, a haplotype derived from *M. domesticus*. This shows a high degree of inbreeding in female lines, as mitochondrial DNA is only inherited through the maternal line. However, the nuclear DNA of these strains shows a higher divergence than would be expected given the degree of inbreeding shown by the mitochondrial DNA. Some Y chromosome characteristics, inherited only through paternal lines, are derived from *M. musculus*, and are shared by house-mouse stocks in Japan and Central China. What this probably shows is that Oriental fancy mice made some contribution to the "old" inbred strains in the early years of developing laboratory-mouse populations.[32]

The breeding, testing, and "sacrificing" of lab mice is a sensitive subject, and this is not the place for a full and balanced treatment. In the UK alone, around 2 million lab mice are involved in experiments every year, though that figure will include many for which "experiment" involves running around a maze or choosing between different foodstuffs. Their very adaptability to a commensal niche makes house mice easy to breed in captivity and to keep at high population densities. Fortuitously, the mouse genome, now fully sequenced, is similar in magnitude to our own, and mice are in many ways a good biological homologue for humans. The welfare of laboratory mice is the subject of regular scrutiny in most developed countries, though they are not covered by the Animal Welfare Act in the United States. Welfare surveys in the UK tend to be reassuring,[33] but stark disagreements arise nonetheless. To some degree, these disagreements arise from fundamental differences of opinion as to what constitutes "good" living conditions for captive mice. Thus Jonathan Balcombe is adamant that conventional lab cages do not provide satisfactory conditions in which to house mice, though his definition of a cage as "a confined space that thwarts basic natural behaviours"[34] rather begs the question of what is "natural" in a highly inbred commensal animal, and rather prejudges the issue. It should be said that Balcombe's paper is repudiated at length and in detail by Robert Blanchard[35] in the same issue of the same journal.

Before moving on to the rats, mention should be made of the spiny mouse of the Mediterranean and Middle East, *Acomys cahirinus*. The common name is a little misleading: it refers to the stiff and rather prominent hairs of the back, and "bristly mouse" would probably be more accurate. Originally a species of savanna and dry scrub across North Africa, spiny mice are found today in free-living and commensal populations, including some in the heart of Cairo.[36] The commensal forms are phenotypically distinct, indicating that this separation of behaviors is of some time depth, and that there is little gene flow between commensal and wild forms. The commensal spiny mice are typically darker in color than their free-living conspecifics, tend to be faster movers and higher jumpers, and show a measurable tendency to greater anxiety.[37] The heightened anxiety is intriguing. One can understand any mammal being anxious in the heart of Cairo, as a natural reaction to the crowds, noise, pollution, and traffic. However, the heightened anxiety state, as measured by glucocorticoid levels in response to stimuli, persists even in lab-bred cohorts of spiny mice originally from commensal populations. Commensalism seems to have selected for a high anxiety response, perhaps because of increased predator pressure in North African towns with abundant feral cats, or because of greater stress induced by the need for spiny mice to aggregate in "unnatural" densities at locations where food is abundant. As we have seen elsewhere in this book, living in dense commensal populations has also affected behavior and aggression in house mice and cats; perhaps it would be wise not to speculate on the behavioral consequences in our own species.

RATS

The success with which ship rats (black rats, roof rats) have spread around the world mirrors the story of house mice. The species probably originates in South Asia, but has accompanied people to all parts of the planet other than the polar regions. Even at the extremes, they come close, having established populations in the Aleutian Islands of the North Pacific and on South Georgia, in the South Atlantic.[38] Ship rat seems to be an originally South Asian species—though, as with many commensal species, the original "wild" range is difficult to reconstruct simply because the species has become so widely distributed. There are "wild" populations of ship rats around the Mediterranean basin today, but these may derive from originally commensal populations that have adapted to a different lifeway.[39] The situation is further complicated by the existence through South and East Asia of *Rattus* species with varying chromosome numbers: 42, 40, or 38. Ship rats normally have 38 chromosome pairs, and the name *Rattus tanezumi* is given to the 42-chromosome group, found from India to Korea and introduced to a number of islands in Indonesia and Micronesia.[40]

Rats appear in the archaeological record of the Middle East around the same time as house mice, on Natufian sites in the Levant. One slight concern is that most of these sites are in, or adjacent to, caves or rock shelters—locations in which archaeological deposits tend to be rocky and open-textured, susceptible to downward movement of small bones. These are also locations at which owls are likely to have roosted, depositing bone-bearing pellets from their predatory activities. In the latter case, rat bones from pellets would show that ship rats were in the region at the time, but not that rats were living commensally with people. The paucity of rats from "open" sites and later Neolithic sites throughout the Levant means that we must treat these Natufian records with caution, always bearing in mind that the later absence of evidence is not evidence of absence, at least not until a great deal more small vertebrate material is published from sites throughout the region. From the Middle East, ship rats clearly spread rather gradually through the Mediterranean region and into Europe, though the record is, again, complicated by questions of stratigraphical integrity and secure dating. A secure record from Slovenia places rats on mainland Europe in the first millennium B.C.E.,[41] though the suspicion is that ship-rat populations were few and far between.

It was the expansion of the Roman Empire that finally gave ship rats both the means of dispersal and the nucleated human settlements that they needed, and by the second century A.D., as we have seen, they had reached Britannia.[42] In that respect, the story of rat dispersal is similar to that of the house mouse, but events at the end of the Roman imperium point to significant differences. Although the actual consequences of Rome's abandonment of its British colony in A.D. 410 are poorly understood, a general decline in urban living is evident from the archaeological record. Buildings were no longer kept up, there is evidence of weathering and soil formation within towns, and the rate of deposition of refuse drops appreciably. From the rats' point of view, the commensal habitat changed to one that favored reduced reliance on human refuse and shelter. Although it is always difficult to *prove* the absence of a species from the record, there is just about enough evidence to suggest that ship rats failed to adapt and went extinct in Britain, while house mice coped more successfully and maintained at least some populations.[43] Assuming that this interpretation of the admittedly insubstantial evidence is correct, it may show not only that ship rats were more closely dependent on the commensal niche than were house mice, but possibly that ship-rat populations were not self-sustaining in Britain. As Roman influence in Britain declined early in the fifth century, so too did the volume of maritime traffic with Continental Europe. Through the second to fourth centuries, rat populations would have been regularly topped up through the inadvertent transport of rats. Equally, one assumes, British rats were carried into Gaul and beyond. With the near-cessation of traffic, Britain was cut off from source populations, and its ship-rat populations withered away.

Maritime traffic resumed in earnest in the ninth century as Scandinavian traders and settlers established new routes and new towns. It is no surprise that ship rats arrived with them, reappearing in the archaeological record in the newly settled trading ports and inland towns.[44] Medieval Europe rapidly acquired populations of rats. Within York, for example, we see ship rats reappear in archaeological deposits from about the end of the ninth century, and they become a fairly predictable presence in suitable depositional contexts (pits, fills, drains, etc.) through to the postmedieval period. Kevin Rielly has reviewed the occurrence of ship rats in British archaeological sites, showing that they are generally common from the eleventh century onwards.[45] The British habit of blaming things on their near neighbors is shown by the fact that both the Welsh and Irish names for "rat" translate as "French mouse"; Rielly infers from this that rats reached Wales in Norman times.[46]

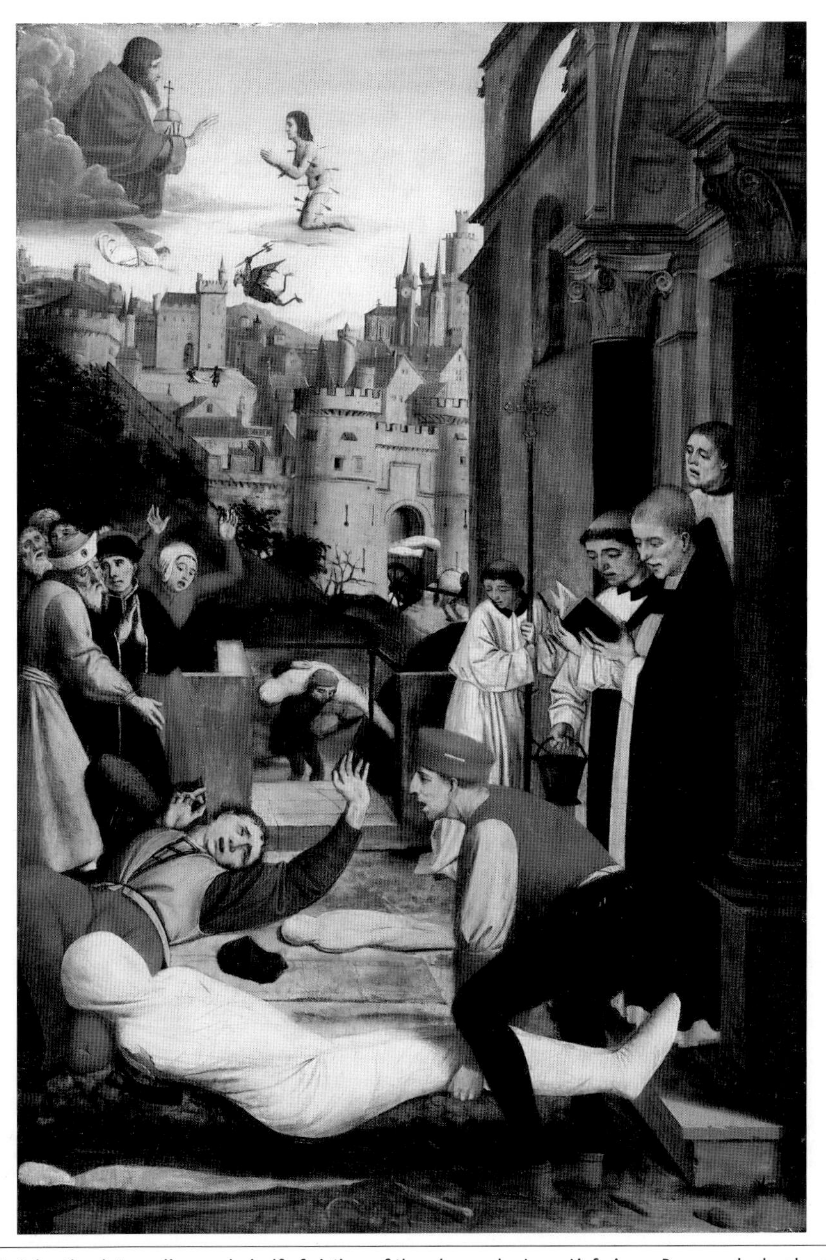

Figure 43. St. Sebastian interceding on behalf of victims of the plague, by Josse Lieferinxe, Provençal school, ca. 1497.

The role of commensal ship rats in the epidemic known as the Black Death is so well-known as to need little elaboration. The rats carried the *Xenopsylla* fleas, the fleas carried the *Yersinia* microorganism, and *Yersinia* caused a fearful outbreak of plague that swept across Europe, devastating England in 1347–51. There has been considerable debate around this familiar tale, including arguments that the historically recorded characteristics of the outbreak do not resemble what would be expected of a plague epidemic.[47] The seasonality of human mortality and the sheer rate of spread of the disease across Europe are said to be more consistent with a highly infectious viral disease. Although it is difficult to find a candidate from among the viral conditions familiar in Europe today, such is the rate of evolution of viruses that the fourteenth-century pathogen need not have a modern representative.[48] That said, analysis of ancient pathogen DNA surviving in human bones from mass graves thought on date and contextual grounds to be "plague pits" from the Black Death has shown the presence of *Yersinia pestis* DNA.[49] The presence of two different strains indicates at least two sources for the outbreak.

Rats reached the New World with Columbus—perhaps literally so, as ship-rat bones have been recovered from a site in Haiti thought to be one of Columbus's camps from his 1493 voyage.[50] Further evocative evidence of rats crossing the Atlantic comes from the Emanuel Point wreck, off Pensacola, Florida. This vessel was part of the fleet led by Tristán da Luna, bringing the first European colonists to Florida in 1559.[51] The ship sank rapidly in a severe storm, taking with it a number of ship rats, the bones of which show poor dental health and a growth defect akin to rickets.[52] Life on board was hard, even for the rats. Enough made it ashore to develop secure populations, and the species evidently moved into Native American settlements quite quickly. The Creek settlement at Fusihatchee, Alabama, yielded several specimens of ship rats from a seventeenth-century phase. Although the majority of bones from this phase were of deer and other hunted resources, just a few chicken bones indicated contact with European settlers, the source of the rats.[53] The replacement of ship rats by common rats is not well documented for North America. In part this is because, as David Landon points out,[54] too few studies in historical-period archaeology differentiate between the, admittedly very similar, bones of the two species. For example, Milne and Crabtree's otherwise fascinating study of early nineteenth-century material from the Five Points neighborhood of Manhattan records plenty of rats, but only as *Rattus* species.[55] At Fort Frederica, on the southeastern coastal plain of Georgia, both species were present in the mid-eighteenth century, carefully differentiated by Elizabeth Reitz, and accompanied by appreciable numbers of bones of cats and raccoons.[56] We need more well-dated and confidently identified records of rats from American historical sites.

To European eyes, sub-Saharan Africa seems to have been unlikely territory for commensal adaptations until relatively recently. Rats are perhaps the obvious commensal animal to look for in Africa: an exotic species that has been closely associated with shipborne trade. Today, of course, African towns and cities have their populations of commensal animals. It would be easy to see this as a phenomenon of the "urbanization" of some parts of Africa, a consequence of the colonial period, and unlikely to date from earlier times. However, just as it can be argued that commensalism may have emerged at least locally in the Upper Paleolithic, so it is at least possible that opportunities for a commensal adaptation were present from time to time and place to place in precolonial Africa. Rats certainly made their homes in early colonial settlements. Melrose House, in Pretoria, South Africa, became the headquarters of the British Imperial forces, and was home to such eminent Victorians as Lord Kitchener.

It is hardly surprising that excavations at Melrose House found remains of rats, among an interesting mixture of "European" food animals (cattle, sheep, pigs, turkeys, chickens, rabbits) and local game (warthog, duiker antelope).[57] More unexpected is the presence of black rats among the bones from Ndondondwane, a major Early Iron Age (ninth century A.D.) site in KwaZulu-Natal. Given the abundance of rodents in the African fauna, identification needs to be cautious: in reporting the Ndondondwane rats, Voigt and von den Driesch point out an earlier misidentification.[58] However, the Ndondondwane specimens included nine skulls and mandibles, making the identifications secure. The most likely source for the rats is the Indian Ocean coast, some 60km from the site as the crow flies, but much further along the twisting Tuqela river valley. A few marine shells from Ndondondwane also point to coastal contact. The site was obviously well connected and may have been of some regional importance, hence an abundance of debris from working elephant and hippopotamus ivory.[59] A later record, though probably indicating similar precolonial contact over a much greater distance between coast and interior, comes from recent work at Ngombezi, on the Pangani River in northern Tanzania.[60] Finds from excavations along the East African coast seem to show rats becoming more common as trading towns developed. Some published records lack precision: small numbers of rodent bones from sixth- to ninth-century levels at Urguja Ukuu on Zanzibar Island could not be confidently identified to species. Positive identifications are mostly later than this in date, from thirteenth- to fourteenth-century levels at Shanga, northern Kenya, and at Chwaka on Pemba Island. Around the fringes of the urban settlement at Shanga, black rats were positively abundant, appearing at the sort of concentrations seen in some European medieval towns.[61]

Figure 44. Muslim architecture in East African Stone Towns is a reminder that parts of Africa had wide cultural and trade links long before the European colonial period. (Source: Ashley Coutu, with permission.)

COMMON RAT

The common rat (brown rat, house rat, Norway rat, lab rat; *Rattus norvegicus*) has a much shorter history as a commensal animal than does the ship rat, but has been even more successful in terms of distribution and abundance. The species probably originates in Asia, perhaps the steppe region of northern China and Mongolia, where it still lives as a burrowing species independently of human habitation.[62] From there, the species spread into central and western Europe, though the timing is a matter of debate. There are apparently well-documented "first appearance dates" for some countries, such as the 1716 date for introduction into Denmark through the port of Copenhagen, where the first appearance of common rats was attributed to the visiting Russian Fleet.[63] To cut an often confusing story short, the species spread rapidly across and around Europe in the eighteenth century, aided to a large degree by shipborne transportation, much as ship rats had been in earlier times. Grzimek has common rats being transported to North America by 1755,[64] another source gives the 1770s,[65] and James Silver says the date of their arrival in the New World is not known, though he does document the first record of the species in the last state to be infested: Montana in 1923.[66]

Common rats are broad-spectrum omnivores, capable of getting some nutritional benefit from practically anything, though showing a preference for meat when it is available. They are intelligent animals, highly social, and capable both of learning individually and of transmitting learned information to other individuals through vocalization and body posture.[67] Common rats can breed year-round, but show an increase in breeding activity during the warmer months. Detailed studies of isolated groups show their phenomenal powers of reproduction. In one Iowa barn, for example, a rat population with negligible immigration rapidly grew over a single year to a peak of one hundred individuals, then declined equally dramatically to just two breeding adults during the following winter.[68] Given that the study was based on intensive, repeated live-trapping, it is possible that the population decline was less extreme, being partly explained by trap avoidance as the resident rats learned what was going on. That said, a more extensive study around the city of Salzburg, Austria, goes some way to confirming the marked seasonal fluctuation in numbers by showing much lower returns of trapped rats and of hairs picked up in hair-collection traps during the winter months.[69] The Salzburg study showed a peak abundance of rats along the riverside (up to 113 rats per kilometer of riverbank), and a much stronger association with water and dense deciduous vegetation than with dry, open places. Interestingly, rats were found in only 35 percent of the seventy-one search locations over 2001–2002, showing that the rats were highly concentrated in favorable locations. This study was undertaken in the context of pest control, which is the motivation for the great majority of published research on the common rat. Nonetheless, the authors conclude that the Salzburg rat population is self-regulating, and that complete eradication is neither possible nor necessary.[70]

The evident success of common rats as commensal animals, and anxiety about their role as disease vectors seem to have limited the research into them simply as an interesting species. What is it about human settlements that common rats find so appealing? In Salzburg, the study concluded that "it seems obvious that the presence of rat can be correlated with the inadequate disposal of waste by human society."[71] A more nuanced analysis was undertaken for England by utilizing the results of the 1996 English House Condition Survey.[72] The results showed that mice are actually far more frequently associated with houses than are rats, and that common rats were noted in 1.6 percent of outdoor areas (e.g., gardens, backyards) and indoors in only 0.23 percent of properties. Both rats and mice were more prevalent where pets or livestock were kept in outside areas, such as dogs in a backyard pen. Perhaps contrary to expectations, both rats and mice were more likely to be found in low-density housing than in high-density housing. What all of this seems to show is that it is primarily food that attracts rats to houses, hence the attraction to dog pens, and that the availability and characteristics of adjacent outdoor space are very important. Rats like older houses because they are more likely to have mature gardens with plenty of vegetation cover, not for any architectural reasons. The presence of cats in the household seemed to make no different to the probability of having rats, showing that adult rats are simply too large a prey for most cats.[73]

Rats certainly act as a reservoir and a vector for a range of significant diseases, including plague and typhus, and have even been identified as a specific hazard for field archaeologists.[74] There is no doubt that they are a hazard to human health, though at least some of the research literature that "talks up" this case originates with colleagues who are employed in the pest-control industry, and there is quite an undercurrent of rat "folk myth" to contend with. One commonly hears statements of the form "In city x you are never more than x yards from a rat." In London, this distance is set at 20 yards (18m), while in Dublin it is a more disturbing "few feet."[75] The Dublin report is the more detailed account, and includes a valuable insight into differentiating the signs left by different rodents, quoting former pest controller Pat Jordan thus: "If you leave a bar of chocolate down and there are bite marks in it, you've a mouse. If the bar is gone, you've a rat." On the one hand, this nicely reflects the common rat's habit of returning food to the nest site for consumption, and on the other hand, it nicely reflects modern urban attitudes to rats.

AND OTHER RATS

In the Pacific region, the past and present distribution and genetics of the kiore *Rattus exulans* has provided a means of tracking human dispersal and colonization across the vast oceanic distances. The first visit or two made by Polynesian peoples to remote islands may have left few recoverable archaeological traces in terms of structures or artifacts. However, the kiore seems to have been introduced in abundance throughout the region, and its remains, and its presence on isolated islands, are eloquent evidence of the human voyages. Introduced populations have been noted on no less than 126 Pacific Ocean islands, and in nearly every case, the population persists today.[76] One of the big advantages of studying this particular species is that it is endemic to Southeast Asia and does not interbreed with the ship rats and common rats that have accompanied Europeans around the Pacific, unlike the pigs, dogs, and chickens that were transported by Polynesians and Europeans alike. Historical records show that the kiore has not been susceptible to transport in recent times, whether because of changes in the form of vessels or cargoes,

Figure 45. Kiore or Maori rat *Rattus exulans*. (Source: Israel Dedham, with permission.)

or because of some behavioral change in kiore or Pacific seafarers. As a result, the populations of this little rat currently scattered across the Pacific are direct descendants of transport during the prehistoric Polynesian expansion across the ocean.[77] Rather nicely, the genetic diversity shown by modern kiore populations is at its lowest on the most remote islands, consistent with the idea that really isolated islands would have been visited comparatively infrequently. The abundant kiore on Rapa Nui (Easter Island), noted by Cook during his visit in 1744, are all of the same mitochondrial DNA haplotype, one that is particularly widespread throughout eastern Polynesia.[78] In fact, the large distribution of genetic haplotypes seems to reflect waves of human colonization. Group I rats are predominantly found in the Philippines, Borneo, and Sulawesi, a region that was colonized by people during the Pleistocene. Group II is distributed through Island Southeast Asia from the Philippines to New Guinea to the Solomons, perhaps indicating one wave of colonization into the western Pacific. Group III kiore spread across the open ocean, with Group IIIa mostly in New Caledonia, Fiji, and Samoa, and IIIb typical of the "Polynesian Triangle," the vast space delineated by Hawai'i, Rapa Nui, and New Zealand, representing the final wave of Polynesian colonization a millennium or so ago.[79]

But did kiore travel as stowaways, as passengers, or as cargo? The species lives as a commensal in many parts of Southeast Asia today, and its adaptability and omnivorous habits make it an excellent candidate for accidental introduction, an abundant commensal which it was very difficult *not* to transport by accident. However, the near-ubiquity and abundance of kiore bones in the refuse middens of early Polynesian sites has led some archaeologists in the region to suggest that they were transported deliberately, as food. In the absence of categorical evidence either way, the use of kiore as food seems highly probable; a small omnivore with, apparently, a particular fondness for coconuts and the capacity to breed numbers rapidly would have been an excellent source of protein en route and also a good species with which to "seed" newly discovered land, to ensure food for subsequent landings. Perhaps the answer lies somewhere in between. Kiore were clearly a highly successful commensal in the past and still are today, and their adoption as on-voyage food may have been a pragmatic way of making use of a species that was already coming along for the ride. Today, they are something of a pest, causing crop damage in coconut and sugar plantations, adding vermin status to their collective CV.

One further rat species deserves a mention. The Australian swamp rat, *Rattus lutreolus*, is not known to adopt a commensal habit with any regularity. The niche is usually occupied, as in most parts of the world, by common rat and ship rat. However, for a few years in the 1970s, a commensal population of swamp rats established themselves in a small zoo at Healesville, near Melbourne, Australia.[80] Like many Australian zoos, Healesville had established populations of common rats and smaller numbers of ship rats. Fluctuations in the numbers of these introduced rats seem to have provided swamp rats with an opportunity. Once established in sufficient numbers, the commensal population of swamp rats was able to persist despite competition from other rat species. Comparison of this commensal population with free-living swamp rats from the same region showed that differences quickly developed. The commensal population had a longer breeding season, the young grew more rapidly, adults were larger, and the commensal rats used their habitat in less restricted ways than their free-living compatriots, showing much less concern about remaining under cover. This is a small but important example, showing that a species that shows little propensity for commensalism may adopt the niche opportunistically if local circumstances are right. As Richard Braithwaite, who studied the Healesville swamp rats, points out, "it is populations, rather than species, that are commensal."

Our relationship with rats and mice extends over 12,000 years and across the surface of the Earth. No other group of vertebrate animals has intruded itself into our lives in quite so many ways, with quite such success, and prompting such a range of human reactions. I have argued that the "companion animal" status of cats arose as people found ways to adapt to and to utilize a commensal species that had clearly moved in with intent to stay. Something of the same transition can be seen in our adoption of house mice as laboratory animals: they are biologically suitable, a useful analogue species, but also they were readily available as a source from which to breed up the many strains now doing service in labs worldwide. Even rats, and of them even the big and forceful common rat, are adopted as pets and spoken of fondly by their keepers. Human fascination with other species is one of the traits that is most characteristic of our species. What rats bring out is the individual variation in that affiliative tendency, as species that are tolerated or even encouraged by one person will trigger fear, loathing, and swift avoidance in another from the same cultural background. No doubt some reaction to, or affiliation with, commensal animals is culturally "learned," but the response to rats shows that there is something inherent in individual variation as well. It is nicely ironic that a species that has been closely involved with behavioral experiments may also serve to bring out our behavioral diversity. Furthermore, there are sometimes hints even in the pest-control literature that murids are admired for their aptitudes and versatility. When the decision was taken to extirpate common rats and house mice from the modest 179 hectares of Motuihe Island, in Watemata Harbour off Auckland, New Zealand, a protracted program of trapping and poisoning was put in place.[81] Eventually, repeated trapping recovered no more rats or mice, and the program was declared a success, but only provisionally: "If a mouse or Norway rat appears again after these five years of absence, there is no way of knowing whether the eradication operation failed or it is a new arrival from one of the many boats that visit the island."[82] Whether we love or loath murids, it seems, we will just have to get used to them.

AND SQUIRRELS

One further rodent merits consideration. The gray squirrel (eastern gray squirrel, *Sciurus carolinensis*) is, as its name indicates, a native of eastern North America, where it is common in woodlands and in residential areas with sufficient cover of mature trees. In the UK, the species is an introduced alien, largely encountered in free-living populations in woodland throughout England and Wales, in which regions it has largely replaced the native red squirrel *Sciurus vulgaris*.[83] In urban parks and leafy residential neighborhoods, gray squirrels have adopted a near-commensal strategy, feeding from food handouts, to a limited extent from discarded organic garbage, and to an increasing degree from food put out to feed native birds. This last behavior has become quite a significant cultural phenomenon among bird lovers in the UK, and we return to it in more detail in a later chapter.

The introduction of gray squirrels to the UK, and their subsequent history, has been detailed by John Long[84] and John Sheail.[85] What that history shows is repeated introduction during the nineteenth and early twentieth centuries, giving rise to numerous isolated populations, then a rapid increase in number and range over just a few decades. The first recorded occurrence of the gray squirrel in the UK was in North Wales in 1828, though the first documented introduction was near Macclesfield in 1876.[86] In subsequent decades, Long estimates

Figure 46. Eastern gray squirrels *Sciurus carolinensis* have adapted very well to urban park and garden habitats in the United States and Europe, including some opportunist changes of diet. (Source: www.nerjarob/nature.com and Mark Willis, with permission.)

there were at least 26 separate introductions at twenty different locations, many of them country parks and estates. The most rapid period of population growth seems to have been the later 1930s and 1940s, the period of the Second World War and its immediate aftermath, during which, perhaps, the spread of an alien species was of little concern or consequence. By the 1950s, numbers had reached such a level that gray squirrels were regarded as a pest, particularly in forestry plantations in which their gnawing of cambium layers detrimentally affects growing trees. Bounties were introduced to encourage their eradication. Between 1953 and 1958, about 1.5 million gray squirrels were killed in England, with little discernible effect on their number or distribution.[87] Within half a century, the charming little mammal that was a desirable addition to one's country park had become vermin.

The role of gray squirrels in the UK today is basically that of villain. Although they continue to be appreciated and fed in urban parks, their pilfering of food provided for birds makes them unpopular with householders, and their place as a reservoir of a virus that is impacting native red squirrels makes them deeply unpopular with conservation biologists. In their urban, commensal populations in the UK and North America, gray squirrels show degrees of behavioral adaptation to the presence of people and of their dogs.[88] In some park populations, squirrels become habituated to taking food from people's hands, and in areas with intensive human activity, squirrels reduce their "alert" distance significantly, ignoring people who are beyond a distance of just a few meters.

Gray squirrel populations in North America and the UK are largely free-living, but the success of urban and residential neighborhood populations shows the adaptability of this species, and its capacity to become a successful commensal. Key to this is a degree of behavioral flexibility, adapting searching and feeding behaviors to acquire food in highly modified environments, and adapting anti-predator strategies to accommodate the higher disturbance but reduced predation risk associated with living in urbanized areas. The native red squirrel has not adapted to the same extent. Isolated populations (and most UK red squirrel populations are isolated) may adapt to feeding around roadside picnic areas or campsites, but red squirrels have not adopted parks and gardens. It is hard to say whether this is simply because the niche has been so successfully adopted by gray squirrels, or whether it shows a lesser degree of behavioral flexibility on the part of red squirrels. Probably both; at any rate, the differing success of two closely related squirrel species is at least in part because one has adopted and exploited the commensal niche with great success.

Of the vast number of rodent species that between them occupy almost every terrestrial biome, just a few have successfully adapted to a commensal existence. Unsurprisingly, it is the most omnivorous of them, the murids, that have done so, though the intelligence and curiosity of these rodents must have been a positive attribute, too. We have a remarkable love-hate relationship with rats and mice, reviling them as vermin while some people quite literally love them as companion animals, and research labs throughout the world value them as nonvoluntary collaborators.

6

Birds

ALTHOUGH RATS AND MICE MAY BE AMONG THE MOST ABUNDANT OF OUR COMMENSAL neighbors, their furtive nature and often nocturnal habits make them less visible. That in turn allows us to forget that they are constantly with us, and to treat the profession of pest controller with a combination of distaste and willful ignorance. Anthropologists speak of the process of "othering" certain groups; pest controllers, like others who deal with the less attractive consequences of our collective lives, would probably recognize that process.[1] Apart from the keepers of rodent pets, most of us "other" rats and mice, setting them aside from our everyday lives, even though we know that they are never far away. However, one group of commensal animals that is difficult to ignore are the birds. They are far more often diurnal in their activities, often occur in large flocks, are sometimes quite noisy, and they arouse in us a wide range of responses and reactions, some of which have changed markedly in recent times. The urban bird par excellence is, of course, the street pigeon, but others deserve a mention because of their abundance; because their association with us has changed in recent years, giving us an insight into process; or because they arouse particular feelings and responses. Time depth is another issue, of course, and we may tend to assume that our close associations with certain bird species are of recent origin. As chapter 3 has shown, that assumption is certainly wrong for some species.

The street pigeon *Columba livia* is a good place to start. The species is so familiar as to need little introduction. Pigeons can be seen in towns and cities on every continent bar Antarctica, and their presence on South Georgia in the 1960s takes them quite close even to the frozen south.[2] They occur right across North America and in urbanized, and therefore more coastal, South America. In Australia, it is the urbanized regions of the east and south of the country that have most pigeon populations, with appreciable populations in the west around Perth. The origins of these widespread commensal populations are discussed later in this chapter. For now, we should note that there are free-living "wild" populations of the same species (hereafter "rock dove") distributed through North Africa and the Mediterranean basin, eastwards through the Middle East, and into southern regions of Central Asia, and that this probably represents the original "natural" range of this cosmopolitan species. In Europe north of the Alps and Pyrenees, it is more difficult to disentangle wild and commensal populations. Apparently wild rock-dove populations persist along the Atlantic coasts of Britain and Ireland, usually in rather low population densities along rocky coasts. That said, pigeons can be seen occupying very much the same habitat and apparently living noncommensally in places on the coast of North Island, New Zealand.[3] These populations can only be derived from introduced, originally commensal stocks that have

Figure 47. Ubiquitous pigeons making good use of medieval tracery on Bury St Edmunds Cathedral; roof in Rethymnon, Crete; and a public square, Gdańsk, Poland. (Source: Bury–Sonia O'Connor, with permission; others–author.)

opportunistically changed niche and behavior. The fact that the British and Irish populations of "wild" rock doves are somewhat isolated to the northwest of the apparent original range of *Columba livia* raises the question of whether these too are former commensals "gone wild," or whether estimation of the original distribution of the species in western Europe is too conservative. The latter explanation seems more probable: the coastal rock doves in Britain and Ireland are far distant from any likely source populations, whereas the New Zealand coastal rock-dove populations are much more plausibly derived from commensal populations in Auckland. There are similar populations in North America, too: for example, in Washington State and British Columbia.[4] The arrival of pigeons in North America goes back to the early seventeenth century, particularly through Canada, presumably as a convenient source of meat.

One of the complications in trying to study street pigeons as a commensal bird is the widespread role of pigeons (as "doves") as a source of food and fertilizer deep into antiquity. The keeping of pigeons was an essential part of the rural estate in the Roman world,[5] with "dovecote" structures identified in surviving remains and in contemporary documents in, among other places, Greco-Roman Egypt[6] and Roman Britain.[7] Recent excavation of a Byzantine-period dovecote in the Negev Desert has shown something of the diet of the pigeons through the evidence of seeds and other plant remains surviving in the accumulation of dung that remained in the dovecote ruins.[8] The remains showed that the pigeons fed freely in the surrounding desert, in fallow fields, and on refuse piles. Some documentary evidence indicates that pigeon-keeping has a deeper antiquity in the Middle East. In the Old Testament book of Isaiah, in a text probably dating from the fifth or sixth century B.C., is the line "Who are these that fly as a cloud, and as the doves to their windows?" (Isaiah 60:8). The allusion to "windows" seems to refer to some form of dovecote, though we should bear in mind the several layers of translation that separate the modern English version from the original. Nonetheless, there seems to be at least the possibility that the provision of housing for free-flying pigeons in order to crop them and/or their dung predates the well-documented Roman practices. Further east, pigeon houses are a significant class of historical structure, with numerous substantial examples, some dating back to the seventeenth century, featuring in the cultural landscapes of Iran[9] and central Anatolia,[10] leaving little doubt of the importance attached to these birds in the Islamic world. That point seems to have been lost on Edward Dixon, a nineteenth-century scholar who wrote at length regarding the keeping of pigeons, and whose love for the birds approached the transcendental: "The gallinaceous birds seem to be representative of the fervid and selfish passions of the East; the Doves to have been created as types almost of Christian virtue."[11] Pausing only to wonder about the "fervid and selfish passions" of chickens, we may note that Dixon's book probably figured in the research reading of that other eminent Victorian pigeon fancier, Charles Darwin. The point is that pigeons qua doves have occupied a significant place in a number of different cultures in Eurasia, ranging from populations that relied upon human feeding and in which breeding was under human control (i.e., they were "domesticated") to populations that were encouraged to congregate in purpose-built structures but left to forage for food and to breed *ad lib.* This latter arrangement meets the definition of "commensal" for our purposes, and is only one step away from those pigeon populations that congregate in structures that were not built for their purposes but which they have adopted. These are the street pigeons that feature in towns and cities worldwide.

Figure 48. Encouraging pigeons. Dovecotes built into barn gable ends in Wharfedale, UK, to encourage pigeons. (Source: author.)

The ecology of street pigeons is quite well-known, though it is sad to reflect that much of the research undertaken on this enterprising bird is directed towards knowing it better in order to contain or eradicate what is often seen as a pest. Social attitudes to pigeons are a topic that we will return to. For the moment, a summary of the ecology of commensal *Columbia livia* may help to show why it has been so successful.

The first point to make about pigeons is that they are enthusiastic and effective breeders. Studies of different populations will obviously give rather varied results according to the conditions in which the pigeons were living. A major study was undertaken at Salford, in northern England, in the years around 1970. At this time, large numbers of pigeons lived, bred, and fed around Salford Docks, then an important point on the far-famed Manchester Ship Canal (and now the northern home of the equally far-famed BBC).[12] The team studied thousands of birds over hundreds of days. Data for breeding success were quite complex in structure and interpretation, but can be briefly summarized as showing that an average pair of pigeons laid between seven and nine clutches of eggs each year, each clutch consisting of two eggs. From those eggs, between six and eleven young per year successfully fledged. Some 75–85 percent of eggs hatched, and about 75 percent of hatchlings grew to fledge. This is quite an impressive rate of reproduction for a medium-sized bird. Unlike most other members of the Columbidae, pigeons can breed year-round. Other species generally have a nonbreeding period to coincide with the winter months. Male pigeons certainly display more often and more intensively during the spring, but breeding takes place in any season. Pigeons are also quite precocious: young birds that leave the nest in January will often breed in the summer of that same year. The Salford study recorded that pigeons tagged while chicks on the nest had successfully paired and were incubating eggs of their own at six months old. Given the opportunity, pigeons are adept at producing a lot more pigeons.

What about food? Pigeons are omnivorous, feeding on seeds, leaves, grain, berries, insects, gastropods, and almost any organic food debris that humans care to leave around. Clearly that is advantageous to the would-be commensal. Pigeons feed in large flocks, which are such a feature of city centers across the world. This behavior raises a number of questions: Is flocking advantageous to individual birds, or does it simply increase competition for food items? Can a flock collectively take better advantage of a feeding opportunity than individual birds? Patient observation of pigeon flocks in a potential feeding space such as a town square or a city park shows something quite interesting. Most of the time, most of the birds are just hanging around. At any given moment, many of the birds in a flock will be actively pecking, but watch one individual bird and it becomes clear that even when food is available, active pecking takes up only a minority of the time spent at a feeding point. Watching the feeding behavior of pigeons entails sitting around in a park, which makes it a popular branch of ecological research. The Salford study logged the behavior of thousands of birds, and their data show that pigeons spent between 66 percent and 84 percent of their "active" time either waiting for food or just resting.[13] This may sound like rather lax behavior, but a pigeon needs only around 30g of food per day, and that quantity can be readily procured in only a small minority of the available daylight hours. For a commensal pigeon, getting enough food is unlikely to be an issue.[14] At least some of that "hanging around" may be related to competition and hierarchy within the flock. Where pigeons have a range of feeding sites, some of which are more productive than others, adult birds will be more abundant at the best sites, juveniles at the suboptimal sites.[15]

Pigeon feeding flocks tend to be quite large. The famous flocks seen in Trafalgar Square, London, formerly reached numbers of around 4,000 birds, though recent moves to ban artificial feeding, and regular harassment by Harris hawks have drastically reduced these numbers to 120–140.[16] Roosting is another matter. Although pigeons can be perceived as a problem because of the numbers in which they roost on buildings, of which more below, their roosting flocks are generally appreciably smaller than their feeding flocks.

Pigeons are characteristically sedentary. Despite the astonishing capacity of racing and homing pigeons to navigate over hundreds of kilometers at impressive speeds, their commensal conspecifics are content to keep to their own neighborhood. In the Salford study, mark and recapture results over one September to February period logged the movements of 793 birds. Of those, 87 percent were recaptured within 100 yards (91m) of their original capture site, and only 5 percent further than 1,100 yards (1km) away. One adventurous bird managed four miles, another three miles. Out of those 793, the only one that had dispersed to any distance was a homing pigeon that had temporarily associated with the commensal flock, been ringed by the observers, then flew to its home loft and a rather puzzled keeper. With classic British understatement, Murton observes that "the species should be regarded as very sedentary under normal circumstances."[17] More recent studies have used tiny GPS receivers to monitor pigeon movements more closely, though with much the same conclusion that they stay close to home.[18] Female birds tend to forage more widely than males, who prefer to keep to abundant and predictable food supplies; any analogy with humans should be avoided at this juncture. One point that emerged from the Basel GPS study was that the details of their observations did not always match with those of other studies beyond the generalizations reported here, and the authors suggest that this indicates a degree of behavioral flexibility on the part of pigeons. Laboratory experiments with pigeons in mazes have also shown a degree of "planning ahead," and seem to indicate that once pigeons become familiar with a spatial environment, they can plan further ahead: two steps ahead rather than one.[19]

Where pigeons choose to feed, roost, and breed depends, of course, on the topography and material composition of the human settlement. Human planning decisions and activities will provide suitable "cliffs" in the form of buildings, and will determine the distribution and reliability of feeding locations. Observations in Amsterdam have shown just how closely the distribution of pigeons maps onto socioeconomic factors in the human population. There are strong positive correlations between the number of pigeons per hectare and the number of people per hectare, the number of houses, and the weight of organic refuse per hectare per annum. In other words, pigeons are most abundant where there are plenty of people, plenty of houses, and plenty of refuse. However, the most intriguing result of the Amsterdam work was a significant negative correlation between pigeons per hectare and average household income. Pigeons, at least Dutch pigeons, prefer poorer neighborhoods.[20] Presumably this reflects a tendency for poorer neighborhoods to have less effective refuse containment and disposal, thus offering more predictable feeding opportunities for pigeons and other commensal species. To digress to another bird species for a moment, house sparrows *Passer domesticus* also show an association with neighborhood prosperity. One factor in their recent decline as an urban bird in the UK is thought to be related to rising affluence: the population decline is much less marked in poorer neighborhoods. Indicators of rising affluence in UK cities include increased vehicle ownership and two-car households, and garden makeovers that commonly feature patios. Unfortunately, the conversion of front gardens to paving for off-road parking, and backyards to patios for burgers and Chardonnay, is not conducive to birds. Furthermore, older buildings are usually better than new houses for nesting purposes.[21]

To return to a point made above, a lot of what we have learned about street pigeons originates in concerns about them as pests. During the author's lifetime, the pigeons of Trafalgar Square, London, have ceased to be an appealing stop on the tourist trail and have instead become feathered vermin, to be actively discouraged by modifying surrounding buildings, minimizing the availability of food, and providing a frequent predator threat by flying tame hawks in the square. The hawks have become something of a sight in themselves, albeit somewhat redundant as peregrine falcons have recolonized central London and other cities and are now trimming pigeon numbers without resort to pest control.[22] It is good to see that the urban food web has acquired a top predator. Reading the literature about pigeons as pests in towns and cities, one senses a degree of overreaction. The control of urban pigeons has become a significant part of the larger pest-control industry, which in 1997 made sales totaling nearly $5 billion in the United States alone.[23] The feeding of birds, predominantly pigeons, in public places is discouraged or actively prohibited. In the UK, it is common to see buildings festooned with netting or fitted with anti-perching spikes—like latter-day *cheveaux de frise*—or daubed with various slippery or otherwise uncomfortable compounds. Because pigeons not unreasonably prefer buildings with plenty of exterior ledges and cavities, it is the elaborate frontages of Victorian Gothic buildings that particularly come in for this treatment rather than the smoother frontages of twentieth-century Modernist or just plain dull civic architecture. But what is the threat?

The following summary of the pigeon problem, and the quotations, are mostly taken from a recent paper on protecting buildings against feral pigeons.[24] First, there is the problem of

Figure 49. Anti-pigeon defenses in Dorchester and Bury St Edmunds. The latter example provides a safe roosting place for feral tennis balls. (Source: author.)

noise: some people find the cooing of pigeons disturbing, and they may even keep some people awake ("Their vocalisations may cause hysteric reactions"—pigeons share this trait with teenage children). Second, and perhaps more seriously, their excreta fouls building façades and adjacent pavements: an individual pigeon can produce around 12kg per annum. Third, pigeons, like most endothermic animals, carry ectoparasites, which may occasionally transfer to people, and pigeons can carry a large number of potential human pathogens.[25] A key word here is "potential," as a recent literature survey showed that only seven of a potential 110 pathogens had transferred to humans, causing 230 cases worldwide, of which 13 cases were fatal.[26] Pigeons are not alone in presenting such a hazard; other wild birds carry human pathogens, which can include serious health hazards such as *Cryptosporidium* and *Giardia*.[27] These two organisms may be transmitted through the water supply, so the contamination of reservoirs by waterfowl can constitute a significant source of intestinal infections. That said, the infection rate among wild birds is typically less than 5 percent. In all, it is sometimes difficult to tell whether the actions that are taken against pigeons are directed towards human health (where the risk is actually rather low), or toward buildings conservation. Given that the protection of civic architecture is not everyone's top priority, the public-health argument, while not completely irrelevant, may be a useful camouflage. The industrial towns of northern England are in the front line of the war against pigeons. Their architecture tends to the externally elaborate. Experimental work shows that pigeons can enter and exit a building through an aperture just 6cm wide with comparative ease, and they will perch in numbers on a ledge of the same width.[28] And the sandstone of which these northern towns are largely built allows a good grip for pigeons' feet, allowing them to perch on ledges of up to a 40° slope.

Netting, slippery ledges, and contraceptive-spiked grain all cost money. Although it has clearly been in the interests of pest-control agencies to stigmatize the pigeon, there is sufficient public acquiescence on both sides of the Atlantic to have tolerated the cost and cityscape implications of restricting these familiar birds. The phrase "rats with wings" has entered common parlance—helped, no doubt, by the newspaper copywriter's eye for a catchy phrase. Colin Jerolmack has analyzed attitudes towards pigeons as a "problem animal," and argues that this attitude can be traced back, at least in the United States, to the middle of the twentieth century.[29] Whether Jerolmack is right to adduce Tom Lehrer's delightful *lied* "Poisoning Pigeons in the Park" as evidence is debatable: it would be far more typical of Lehrer to have chosen and perverted an activity—feeding pigeons—that was still regarded as acceptable and harmless.[30] Nonetheless, Jerolmack sets out the deliberate association of pigeons with disease in 1960s New York, showing how the "rats with wings" metaphor served to change attitudes—first in civic authorities, then increasingly among the general public. However, understanding the historical trajectory still leaves open the initial question: why? Public health looks more and more like a post hoc justification. Dog feces are far more abundant than those of pigeons and constitute a far more prevalent health hazard, and the consequences of dog bites put pressure on public-health systems.[31] However, dogs are seen as domestic animals, animals of the hearth and home, not opportunists that have moved into our towns from "the wild," even though that description might equally well have applied to dogs early in their association with people. Jerolmack and others have argued that human societies, at least in the industrialised West, require a distinction between nature and culture. Towns are culture, pigeons are nature, and the impudent ease with which the latter have invaded the former is culturally unacceptable. The problem with pigeon feces is thus a pollution that is as much symbolic as it is actual. Similar issues arise with some other commensal species, and we return to them later in this volume.

Figure 50. Jackdaws *Corvus monedula* gathering at dusk to roost in the old city of Gdańsk, Poland. (Source: author.)

There are other urban birds, of course, some of which may be commensal, within the terms of this volume. House sparrows have already been mentioned, another creature of the "wild" that has made itself very much at home. The modern distribution of this species matches that of the pigeon, making it difficult to determine its "natural" distribution. As we have seen in chapter 3, bones of sparrows appear in archaeological assemblages in the Middle East earlier than in other parts of Eurasia, though it is difficult to be certain that this indicates their region of origin or simply the region of the world that first developed the nucleated settlement into which sparrows invited themselves. Sparrow bones are small and easily destroyed or overlooked, so it is not surprising that their record is sparse. There is some evidence of a spread from the Mediterranean region through Iberia in later prehistory, perhaps following a similar route at a similar time to house mice.[32] Once established in the more temperate parts of Eurasia, sparrows accompanied people into the more northerly and offshore regions, and it is those movements that have been recent enough to have featured in historical records. Like so many things European, sparrows entered the United States through New York in the mid-nineteenth century. Some authors record that the first introductions were an effort at biological control of pests: the sparrows were intended to control cankerworm infestations.[33] Others acknowledge this introduction, but point out the habit of migrants of carrying something familiar from "the old country," arguing that appreciable numbers of sparrows probably entered the United States and Canada as caged birds accompanying European immigrants.[34]

Further releases placed sparrows on the West Coast, and the species is estimated to have colonized much of North America by the early twentieth century. Similarly, the present-day Australian population derives from deliberate introductions in the second half of the nineteenth century.[35] One estimate has it that of thirty-nine known deliberate introductions of house sparrows, thirty-three have successfully established populations.[36]

But are sparrows fully commensal, in the sense of sharing food sources with people, or are they just synanthropic, in the sense of choosing to live alongside us? The early association of people and sparrows in the Neolithic Middle East seems to have been centered around cereal cultivation—indicating, perhaps, that sparrows were attracted by the storage and spillage of grain as an easy feeding option. This association has been used to explain the spread of sparrows into northern Europe, possibly as a side effect of the widespread adoption of horses for riding and haulage and the consequent creation of habitat patches rich in fallen grain.[37] The recent decline in house-sparrow numbers in the UK has focused attention on what it is that sparrows actually need from human settlements. Several surveys have shown the importance of what house-proud humans might regard as "untidy" gardens and allotments.[38] This may be a combination of things: overgrown corners provide shelter; seed heads are likely to be left on plants, providing food after the summer; and those overgrown patches probably harbor a high invertebrate population. In this model, sparrows are not deriving food directly from our food, but they are deriving feeding opportunities from our highly modified environment. That is perhaps subtly different from the feeding behavior of pigeons, reminding us of the many different forms of association that can exist between ourselves and other species.

Starlings *Sturnus vulgaris* provide another subtly different association, one that is perhaps not truly commensal despite the large numbers of starlings that occupy our city centers. Whereas pigeons form large feeding flocks in towns and cities, their roosting flocks tend to be much smaller; with starlings, the reverse is the case. The typical daily movement of starlings in an urbanized environment is to disperse into the surrounding suburbs and countryside during the day—feeding on invertebrates in particular, plucked from pastures, lawns, and sports fields—and to return to town centers in the evening in huge roosts. These roosts, which may amount to thousands of birds, are matched in the countryside by similarly sized flocks that adopt, for example, a reed bed as a roost site. Starlings have adapted their roosting behavior to take advantage of relatively protected places in towns and cities, gaining some warmth and security, but their feeding behavior remains largely noncommensal. In this they contrast rather well with pigeons. Nonetheless, urban starling roosts are often regarded as a problem. The first major survey of urban roosts in Great Britain was undertaken in the early 1960s, initially because of the perception that the habit was becoming common.[39] It is clear from the opening paragraph of Potts's 1967 article on urban starlings that fouling and a perceived disease hazard were seen as reasons to discourage starling roosts in towns. A more recent survey has highlighted the flexibility that starlings show in targeting roost sites, and the authors propose that this flexibility could be exploited in order to encourage the birds to roost elsewhere.[40] One thing that may act in favor of urban starlings is the behavior of flocks immediately prior to settling for the night. Tight flocks of numerous birds soar and swirl against the evening sky in close formation, creating a remarkable spectacle that is widely appreciated. Although the best of these evening flights are to be seen in open countryside, over reed beds and similar roosting sites, far more people see the urban flights.

A species that loudly demands our attention is the ring-necked or rose-ringed parakeet, *Psittacula krameri*. Originally a species of tropical Africa and South Asia, these birds have become a familiar site in towns and cities in the United States and, especially, the UK. There is a charming myth to the effect that their UK population owes its origins to a pair of birds released from captivity in London by no less than Jimi Hendrix. The dates seem to match: Hendrix was based in London as his fame grew in the late 1960s, and the first record of a family group of the parakeets was recorded in 1969. Another fondly held story has it that the population derives from a 1951 escape of numerous parakeets at Shepperton Studios, on the outskirts of London, where the birds were acting as "extras" in the film *The African Queen*.[41] Sadly for both myths, though not for the birds, rose-ringed parakeets had briefly established themselves on the outskirts of London in the 1930s, and the first report of them in the UK dates from 1855, in Norfolk.[42] Following the 1969 sighting, the parakeets established themselves successfully around London and other locations, mostly in the Southeast of England, and were formally accepted as a "Category C" species (an "established exotic," not unlike Hendrix?) in 1983.

Figure 51. The rose-ringed parakeet *Psittacula krameri* is now abundant in many urban locations. (Source: Dennis Jarvis, Wikimedia Commons.)

By the end of the twentieth century, the UK population of rose-ringed parakeets was numbered in the thousands, with one roost alone estimated at 2,500 birds.[43] Perhaps inevitably, the novelty had begun to wear off, and people living near to these roosts began to grumble about the noise, the parakeets' vocalizations sometimes causing hysteric reactions. More to the point, concern was expressed about the impact that the new species would have on native hole-nesting species. There is, in fact, precious little systematic published research on this latter point. A study in Belgium sought to assess the impact that rose-ringed parakeets might have on native populations of nuthatches *Sitta europaea*. Nuthatches are small, hole-nesting birds likely to be affected by competition from a larger species. The Belgian study concluded that the competitive impact was limited, noting that the parakeets tended to be most abundant in rather fragmented woodland on urban margins, whereas nuthatches were most abundant in more continuous stands of old woodland.[44] Parakeets are also accused of outcompeting native species at garden feeders, again on the basis of little systematic research and despite some evidence to the contrary.[45] In just a few decades, the parakeets have achieved the trajectory that we have seen for pigeons, from neighbor to pest. In 2007, the statutory body English Nature announced that it would relax the protection given to rose-ringed parakeets under the 1981 Wildlife and Countryside Act, permitting them to be culled without license where they constitute a pest or hazard. Public reaction, judging by blogs and vox-pop interviews, is still somewhat mixed. Some speak up in the birds' defense, describing them as a vivid and appealing addition to "our" wildlife, while others complain about the noise and alleged impact on native species. Particularly relevant here are the complaints that the parakeets take peanuts and other food that has been put out for other birds; we welcome commensal birds into our parks and gardens, but only on our terms, it seems. In fact, it hardly seems correct to describe these parakeets as commensal. Their behavior is much more akin to that of starlings: forming large roosts on the fringes of towns and in open urban spaces, often exploiting isolated patches of tall trees, then mostly dispersing into adjacent countryside to feed, with occasional forays into gardens.

These few bird species reflect something of the difficulty of categorizing the relationships between ourselves and our many neighbors. The pigeons that flock in our squares and parks are indisputably commensal, exploiting our food wastes and spillages to good effect, and engendering responses that range from affection to outright hostility. Sparrows probably slip into the "commensal" category, though it is the feeding opportunities in the modified environments that we create that matter as much as any direct benefit from human food crops or waste. Starlings and rose-ringed parakeets are both synanthropic—starlings perhaps the more so—but both rely on people more for roosting opportunities than for food. It remains to be seen whether rose-ringed parakeets will adapt to regular use of garden bird feeders, becoming more strictly commensal, or whether competition from other species already well adapted to this niche, and direct action by members of the public will inhibit this adaptation.

Gulls, *Larus* spp., have become a more and more familiar sight around towns and cities in Europe over the last few decades, utilizing our structures for the purposes of nesting sites (flat roofs) and food (garbage landfill sites). Again, the gulls have made themselves unwelcome, disparaged in one recent survey for their "noisy and aggressive nature."[46] In Britain, their adoption of an urban, commensal habit is often attributed to the 1956 Clean Air Act, which legislation effectively brought to an end the large-scale incineration of refuse, and hence greatly increased the amount of organic detritus that went to landfill sites.[47] In fact, urban nest sites had been in use since the early 1940s. The 1960s saw a substantial increase in urban gull numbers, probably because of food availability, but that was not the initial factor. A trend towards flat roofs in decades either side of World War II may have been the factor that first gave gulls their foothold. Certainly, flat or gently sloping roofs are preferred today, ideally those that are old enough to have developed growths of moss and lichen. Projecting bolts and other impedimenta are useful to help to anchor nests, and nests tucked among groups of chimneys are not an unusual sight. The species involved in northern Europe are usually herring gulls *Larus argentatus* and the closely related lesser black-backed gull *L. fuscus*, replaced in southern Europe by yellow-legged gulls *L. michaehallis*. Urban gull breeding success is enhanced over their "wild" neighbors, with breeding starting earlier in the spring, less predation of young, and a longer feeding day facilitated by streetlights.[48] The role of food as a simple factor is questionable. A detailed comparison of breeding success by yellow-legged gulls in Venice compared with "wild" colonies on the adjacent lagoon concluded that food availability was not a major factor in the success of the urban birds, but reduced predation risk was important.[49]

Black-headed gulls *Chroicocephalus ridibundus* have also found towns to their liking, particularly in winter. A study of winter roosts in Krakow, Poland, showed the gulls to have a distinct preference for flat roofs, a common feature of postwar Polish architecture, but also showed the strongest correlation with the distribution of litter bins.[50] As usual, the gulls are described as a "nuisance." An interesting adaptation on the part of black-headed gulls is their habit of kleptoparasitizing pigeons and corvids: commensalism by proxy? The adoption of urban habitats by black-headed gulls is usually regarded as of quite recent origin, as with other gulls, though the recent find of a black-headed gull skeleton in a seventeenth-century context in York may be an example of earlier adaptation on the part of this versatile species.

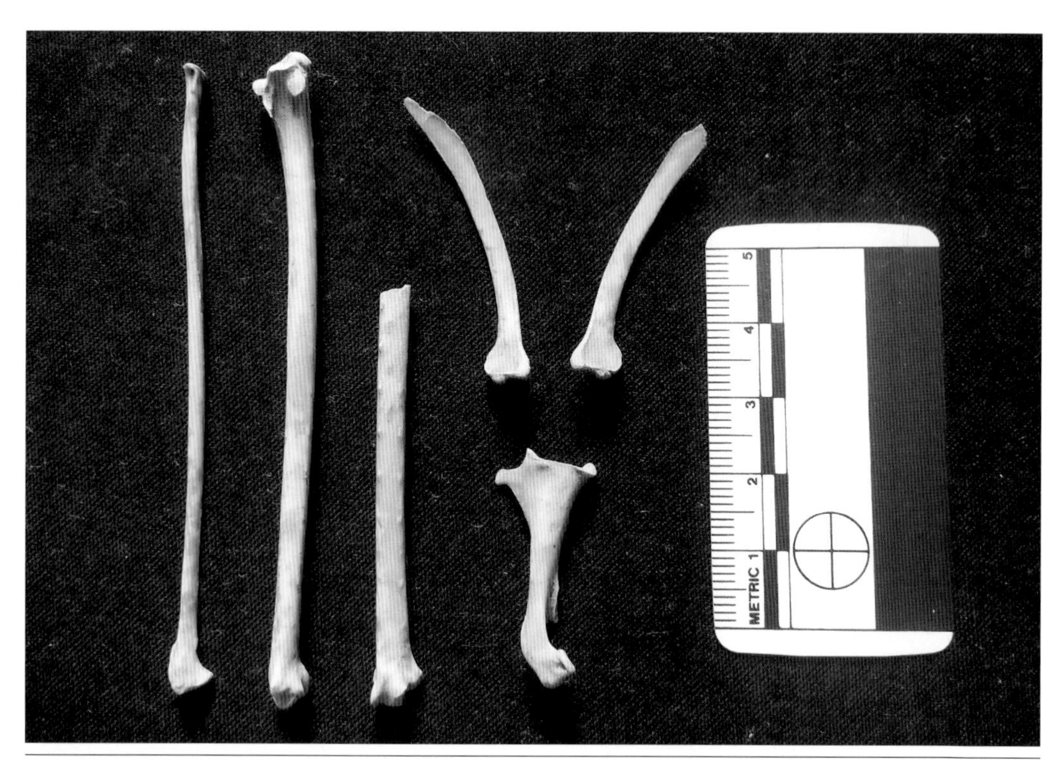

Figure 52. Black-headed gull *Chroicocephalus ridibundus* from excavations in York could possibly been early commensal adapter. (Source: author.)

One more family of birds deserves discussion here. The crow family, the Corvidae, are near-global in their distribution, and have a well-earned reputation as intelligent and behaviorally flexible omnivores. The question of intelligence has been explored in some depth for the corvids, with a number of authors arguing that corvids generally show cognitive abilities similar to those of primates and distinctly different from those of most other bird families.[51] The best-known corvid tool-users are New Caledonia crows *Corvus moneduloides*, which regularly make tools of a highly specific functional form out of twigs and *Pandanus* leaves, though other corvids have shown remarkable tool-making capacity in experimental situations.[52] It has been proposed that ravens *Corvus corax* show a "theory of mind," the capacity to consider what another individual may be thinking in a given situation and to adapt their own behavior accordingly.[53] In particular, observation of food caching and pilfering among a group of ravens showed that individual birds could deceptively manipulate the behavior of others in order to increase their pilfering success. Sad to say, the detailed description of crows learning to operate a coin-operated vending machine turned out to be a highly plausible hoax.[54] However, while observing jackdaws with the aim of acquiring images for this book, I watched an adult jackdaw "lie" to a persistently begging younger bird. The adult was soliciting pieces of bread from picnicking humans while walking back and forth along a low wall. At one point, having been repeatedly hassled by its fully grown offspring, the adult bird kicked a piece of bread off the wall, out of sight of the immature bird. That might have been accidental, but every time the adult walked past a point immediately above the bread, it quickly glanced down and walked on. Eventually the immature bird lost interest, whereupon the adult hopped down to the "concealed" bread and flew off with it. The two of us who watched this small drama were left in no doubt that the adult bird had quite deliberately concealed food from the other, then behaved so as to conceal the fact that it knew where there was some food.

This intelligent behavior leads many corvid species to be highly successful commensals. Ravens commonly follow packs of wolves in order to scavenge the remains of successful kills, and seem to have adapted this behavior to such a degree that they will converge on the sound of rifle fire in wooded areas in order to scavenge the kills of human hunters.[55] They do not show the same behavior in open terrain, possibly because the ravens do not need the audible signal in terrain where feeding opportunities are more visible, but possibly also because they are much more vulnerable in an open landscape than in woodland. Their response is thus not simply a learned response to a single stimulus (gunfire), but something more subtle and flexible that depends upon the stimulus and the context in which it is heard. In general terms, human relations with corvids have often been somewhat ambiguous. In North America, the role of American crow *Corvus brachyrhynchos* as a pest in cornfields has only declined relatively recently as the species has taken to roosting in urban areas and exploiting human food waste.[56] In India, the house crow *Corvus splendens* is a long-standing and ubiquitous commensal, highly regarded in popular culture for its versatility, even leading to the suggestion that the epithet *splendens* "presumably refers to its splendid impudence, for the house crow is forever harrying and swearing at people, kites and dogs and bullying smaller birds."[57] In fact, *splendens* is usually translated to mean "brilliant" as in shining, which is something of a pity. Our admiration for corvids is seen in the way that stories of "crow intelligence" proliferate, whether genuine (crows placing walnuts on roads so that cars will crack them open) or not (Klein's vending machine). John Marzluff has argued that ethological adaptations of crows to people and of people to crows constitute a cultural coevolution, a perspective that may be applicable to more species than crows, and to which we return in a later chapter.[58] However explicitly we model this relationship, there is no doubt that people and corvids have shared living space and (probably) food for millennia. Like mice, corvids appear in the archaeological record of some of our earliest nucleated settlements.[59] Even in a culture without villages, there are hints of a close association. Jon Driver used the particular deposition of two raven skeletons at the Paleoindian site of Charlie Lake Cave, in western Canada, to suggest that the ravens were more than a mere chance occurrence, citing the important symbolic place of ravens in later Native American cultures.[60] Another possibility is that some ravens had already learned the advantages of living alongside people, reducing their own foraging time by waiting for the human neighbors to bring in food, a share of which could be scavenged by the ravens from the inevitable garbage. In that model, the ravens had culturally incorporated people, even though the people had not yet culturally incorporated the ravens.

We see corvids throughout the archaeological record of human settlements—seldom in great abundance, rarely in anything other than the garbage pits and dumps, but a consistent presence nonetheless. In the city of York, UK, ravens, crows, and jackdaws have been present from Roman times onwards, the most regularly encountered species after the domestic ducks, geese, and chickens. By medieval times, the raven is the most regularly recorded of the three corvid species,[61] though it is also the largest and therefore the most conspicuous during excavation, making it a little more likely to be recovered than, for example, the appreciably smaller jackdaw. Nonetheless, the absence of ravens from York and its surroundings today contrasts with the impression that we gain of their near ubiquity in the medieval period. Some change in the environment of the city or in the behavior of people or of ravens has done away with the commensal niche that ravens successfully utilized for some 1,500 years. Of course, the city has grown in that time, though even today York is only a few minutes' flight time in diameter. More likely it is changes in the disposal of refuse, with less surface accumulation of organic garbage, that has affected the niche, perhaps compounded by a reduction in mature trees in which ravens could nest close enough to the city to maintain a breeding population in the locality. The red kite *Milvus milvus* was also present within the city through to late medieval times, then vanishes from the record,[62] and a similar explanation may be advanced. Unlike that of ravens, the drastic reduction in kite numbers and distribution within the UK became a matter of concern to conservation bodies, who successfully used archaeological evidence that the species had formerly been widespread and numerous to argue for a program of reintroduction.[63] Although kites have not yet recolonized the city, they are now a common sight in the surrounding countryside, benefiting from the reintroduction program and probably sustained largely by the abundance of rabbits, another human introduction from medieval times. The cultural coevolution of kites and people in the UK has worked to the kites' advantage, though the same cannot be said for ravens.

It is easy to sum up the successful commensalism of corvids worldwide by simply pointing to the feeding opportunities that people deliberately or inadvertently provide, though this may be too simple. In a detailed study of crows in two cities in southern Finland, Timo Vuorisalo and colleagues seem almost to have demonstrated the opposite.[64] Crows only moved into Finnish cities in the early part of the twentieth century, and then remained uncommon until the 1960s, so their change in number and distribution has occurred during the period of detailed biological recording. The establishment of large urban parks late in the nineteenth century may have been the stimulus for the initial colonization. However, the availability of organic garbage probably reduced rather than increased through the twentieth century, so the late increase in numbers is not correlated with access to food. In Turku, for example,

Figure 53. Corvid behavior is demonstrated by a house crow *Corvus splendens* in a café in Dar es-Salaam, Tanzania (left), and jackdaw *C. monedula* in a café in Ambleside, UK. (Source: jackdaw—Sonia O'Connor, with permission; house crow—author.)

crow numbers increased only after the construction of a municipal garbage incinerator greatly reduced the organic waste available in landfill.[65] Other factors may have been more significant, among them a reduction in predation of nests and young birds, particularly by small boys. Increased tolerance and protection of urban wildlife could be seen as a cultural change among the human populations, while crows have developed a greater tolerance of people, shown by greatly reduced flight distances.[66] In short, the story of crows in Turku and Helsinki through the twentieth century nicely illustrates just the cultural coevolution proposed by Marzluff and Angell. No one supposes that the reduced flight distance of crows or the increased tolerance by people has a genetic basis—Darwinian natural selection cannot claim these adaptations—yet both have spread through the respective populations by imitation and learning, to mutual benefit.

Birds are among the most obvious and familiar of commensal animals, sometimes seen as vermin, but sometimes actively encouraged by artificial feeding and a high degree of tolerance. As a group, they illustrate the range and subtle variations of neighborly living, with a complete gradation from the fully domesticated to wholly commensal, to synanthropic but not commensal, to wholly free-living and "wild," potentially seen in different populations of pigeons alone. And being diurnal and often quite noisy, birds force themselves on our attention. Much of this chapter has been written to the accompaniment of an intermittent barrage of harsh squawks from jackdaws who live their noisy, social lives on and around nearby trees and rooftops, and who never, ever miss the chance of a free meal.

7

Commensalism, Coevolution, and Culture

PREVIOUS CHAPTERS HAVE SOUGHT TO EXPLORE THE TIME DEPTH OF COMMENSALISM, and have looked at some of the many species that have adapted to our living space. Now we draw together and consider some of the themes that have emerged from this survey, to discuss commensalism as a strategy, to consider our responses to it, and to think about the future options for ourselves and our neighbors.

The introduction to this book considers definitions of terms, offering largely functional definitions ("Animals are synanthropic if they live their lives in *this* way . . .") in order to provide a satisfactory working vocabulary for the subject in hand. Although such definitions are useful and reasonably straightforward, we have also seen the importance of the human response to the other species, the cultural factors that differentiate welcome commensals from vermin. A vocabulary that is satisfactory from a biological perspective may not be so from an anthropological one. Another means of definition is by reduction: we can reduce the complex past and present of commensalism by unpacking the topic into its key defining components.

First, extent. The adoption of a commensal niche by some animal species seems to be genuinely global. Where we go, they are. To some extent this reflects the interconnectedness of human populations. We have seen how smaller commensal rodents have closely reflected movements of people through the Mediterranean and western Europe, and around the far-flung islands of the Pacific Ocean. Cats have accompanied us everywhere and have sometimes persisted where we have not, and pigeons are a feature of urban living just about everywhere. The global spread of these few species in our wake is a major factor in what McKinney calls "biotic homogenization," through a process that others have termed *ethnophoresy*.[1] However, there is local recruitment too. Monkeys in India, cuscus and civets in Southeast Asia, raccoons in North America: different parts of the world have their characteristic commensals. What that tells us is that it is the niche that matters, not the habitat, the faunal community, or the biome. That conclusion has, in turn, implications for when and where this behavioral strategy may first have manifested.

Second, chronology. Although the evidence becomes patchier and more debatable further into the past, commensalism extends at least to the beginning of the current postglacial climate phase and, I would argue, further back into the Late Pleistocene as faunal communities adapted to increasingly abundant and intrusive Upper Paleolithic peoples. Further investigation of that time depth would require a more detailed examination of the Paleolithic record. The broad overview given here can only hope to establish the prerequisite of at least allowing the possibility of commensalism in the Paleolithic and therefore subjecting it to more detailed critical scrutiny. Even in a cautious analysis, we can see that commensalism did not require towns and cities, not even villages. All it required was patches of human activity of sufficient intensity to perturb and disequilibriate preexisting communities, altering patterns of predation and competition, and generating a trophically rich opportunity. Given those circumstances, however, wherever and whenever generated, the early-adopter commensals—mice, foxes, corvids—moved in.

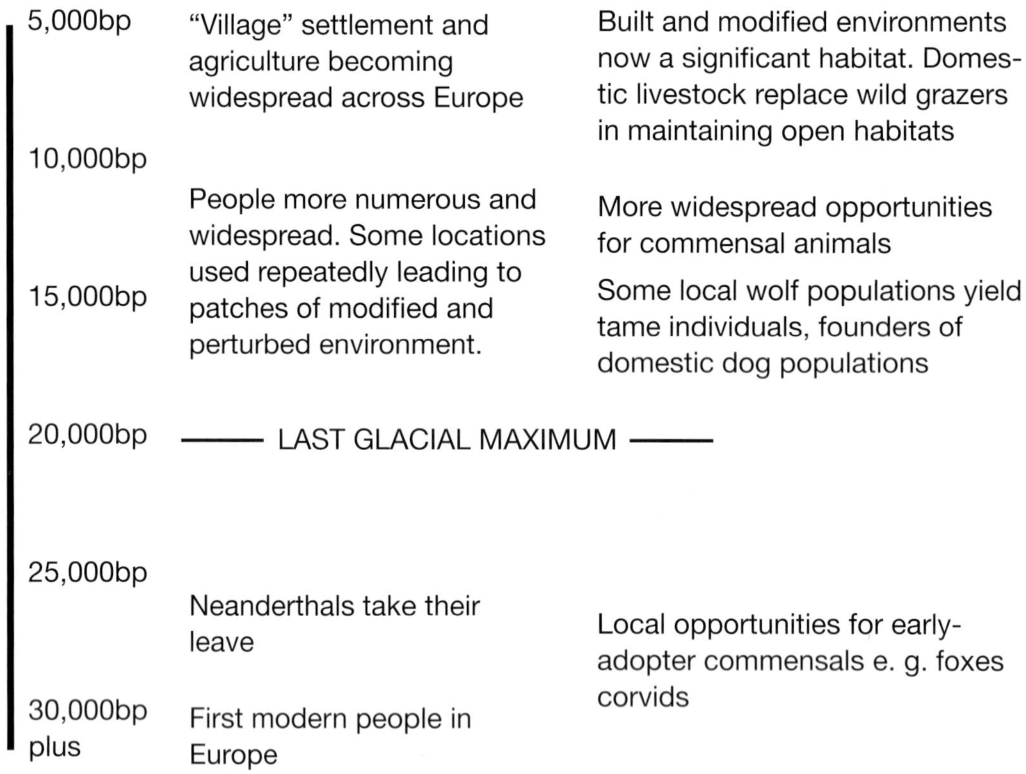

5,000bp	"Village" settlement and agriculture becoming widespread across Europe	Built and modified environments now a significant habitat. Domestic livestock replace wild grazers in maintaining open habitats
10,000bp		
	People more numerous and widespread. Some locations used repeatedly leading to patches of modified and perturbed environment.	More widespread opportunities for commensal animals
15,000bp		Some local wolf populations yield tame individuals, founders of domestic dog populations
20,000bp	—— LAST GLACIAL MAXIMUM ——	
25,000bp		
	Neanderthals take their leave	Local opportunities for early-adopter commensals e. g. foxes corvids
30,000bp plus	First modern people in Europe	

Figure 54. Possible timeline for the emergence of commensalism in Europe.

Third, scale. Just because some fox populations may have become commensal in Europe during the Upper Paleolithic, that has no implications for the status of *Vulpes vulpes* as a species any more than the thoroughly urban adaptation of some UK fox populations today says anything about the ecology or ethology of their rural conspecifics.[2] Commensal adaptation is at the level of population, not of species. Over much of their global range, it is true, pigeons live commensally, but southern Europe has "wild" free-living populations, and New Zealand even has wild populations that derive from commensal ancestors.[3]

Thus we have an adaptation that is population-specific, not species-specific, and neither culture- nor biome-specific. If the heart of the matter lies in the interaction of a particular animal population with a particular human population, then we are dealing with a *cultural* phenomenon as much as a biological one. We have seen that human responses to commensal animals vary considerably between cultural groups, and through time within any one group. Within my own lifetime, attitudes to London pigeons have passed from willing acceptance as "local color" to outright persecution. The same trajectory has been identified for a number of other species. Clearly this is not the result of some genetic drift in the human population of London, increasing an "intolerance" allele and bringing about a collective change of attitude within just two generations. *Opinions* have changed, for whatever reason, and those changed opinions have been promulgated. In short, attitudes to commensal animals are learned, and that is the essence of a cultural phenomenon—one that is population-specific and transmitted by learning, whether vertically between generations or horizontally between peers.[4] Commensal animals have been a cultural influence on people from the Upper Paleolithic onwards.

In some instances, our cultural response to our neighbors has been essentially utilitarian. The foxes that were attracted to Upper Paleolithic and Natufian sites were occasionally captured and butchered, presumably for their meat and fur.[5] Seen from the human side of things, this relationship must have resembled "garden hunting," by which horticultural peoples procure meat and other resources by hunting the animals that are attracted to their cultivated plots. For foxes, the enhanced mortality did not outweigh the benefits of access to food, just as the high mortality of urban foxes today has been no barrier to their adoption of our towns and cities.[6] Perhaps guinea pigs fall into the same category: a species that took advantage of the availability of food, and was in turn utilized for food. In the Pacific region, we see the kiore in the same light: a species that was not necessarily initially adopted as a source of fresh meat by a mobile maritime people, but one that lent itself to that role, enabling the Polynesians to derive benefit from a species that already derived benefit from them.

There are other forms of utility. The cats of the Middle East and civets of Southeast Asia may have been attracted by the feeding possibilities of human refuse, but it seems likely that rodent prey was also part of the attraction. That would have made these species distinctly useful, ensuring that they were tolerated. In the case of cats, their adoption into our homes and lives has been deep and lasting. There is a nice parallel to be seen in the increasing numbers of peregrine falcons that inhabit our major cities: our refuse attracts their prey, and our buildings provide their perching and nesting space.[7]

What all of these species have in common—foxes, cats, kiore, and more—is that certain of their populations learned how to gain benefit from associating with humans, and those humans learned ways of gaining benefit in return. What we have, therefore, is a cultural coevolution, a mutual process of learning to live together that has gone on sometimes quite locally, and sometimes quite temporarily. The implication of this view is that the process has been a cultural one *on both sides*, and therefore that the animal species concerned must be regarded as cultural entities.

Debates over the cultural status of other species tend to focus on material culture, notably the capacity of other primates to utilize, modify, or fabricate tools of some form.[8] The ability of some groups of chimpanzees to make and use tools is well-known and needs little elaboration.[9] Given that ability in chimps, and something similar in capuchin monkeys[10] and in crows,[11] it seems perverse to deny that some populations of some other animal species have material culture, the key defining attributes being that the tools are population-specific, and that their manufacture and use is learned. However, tools are not the only manifestation of culture. Macaque monkeys in northern Japan make use of warm springs to maintain their body heat during winter, a habit that is not general to the species and to which youngsters have to be introduced.[12] The warm springs are, in effect, tools that are used for their survival value, an extrasomatic means of adapting to a seasonally rigorous climate. Commensalism falls into the same category: a cultural adaptation of behavior by which some populations of some species gain benefit, whether by extension of range, increased population density, reduction of competition, or resource buffering against hard times.[13] And in parallel, the sympatric human populations undergo a cultural adaptation that may be utilitarian, as discussed above, or actively antipathetic or, perhaps, indifferent.

A quite fascinating example of that coevolution links the introduction of gray squirrels into Britain with the habit in that zoophilic culture of putting out food for wild birds. When I was a small child in the late 1950s, it was common practice to put out stale bread crusts or biscuits for the garden birds to eat. They were fed, but only on scraps that were no longer palatable to people. I still recall my mother's horror on finding that my father had cut up several slices of perfectly good bread in order to feed the birds during the memorably cold winter of 1963. As with so many aspects of life, feeding the birds became commercialized, and specialist suppliers began to sell not only food intended for "wild" birds, but also a range of devices in which to provide the food. Squirrels quickly discovered that peanuts and sunflower seeds, while not normally available in the British countryside, are nonetheless palatable, especially when available in large concentrations. Food put out for finches (Fringillidae) and tits (chickadees, *Parus* spp.) was taken by squirrels with increasing frequency. The habit of young squirrels staying with adults for a period of time after leaving the drey made it almost inevitable that this behavior would be learned and rapidly transmitted through squirrel populations. The response of many bird-feeding people was to deter the squirrels. Hanging the seed containers from lengths of string or wire made it more difficult for squirrels to access them, but squirrels are adept climbers and that practice was soon inadequate. Seed containers were adapted to be less accessible, for example by adding a curved, overhanging lid, and then by adding an outer cage designed to allow access to small birds while excluding squirrels. In short, squirrel populations have undergone changes in behavioral culture, and people have undertaken changes of material culture in response and in parallel. There is a thesis to be written on the typology of birdfeeders. More to the point, we may reasonably ask why it is that so many people are prepared to spend good money providing supplementary feeding for avian commensals, yet strongly resent it when that same food is taken by a mammalian commensal. It may be nothing more than the fact that gray squirrels are not native to Britain: zoophilia tempered by xenophobia, also shown by our reluctance to feed rose-ringed parakeets.

The birds and squirrels example illustrates both zoophilic and antipathetic reactions. Genuine indifference may be rare, even exceptional. Just because we neither directly utilize nor seek to exterminate an animal neighbor does not mean that we are disinterested in it. This book—indeed the whole series of which it is a part—is a reminder of that fascination with which people regard other animals. We eat, fear, anthropomorphize, and worship them, endowing animals with attributes that they do not necessarily have, and concerning ourselves with species that are of no conceivable utility. Concern over the decline of house-sparrow populations in the UK is a good example: we make no use of them, they are of no benefit to us, yet we are concerned to understand and arrest their decline.[14] Crucially, however, this zoophilia is not a cultural phenomenon. It is species-wide, something fundamentally characteristic of human populations throughout the world, throughout recorded history, and probably for much of our prehistory. Children do not have to be taught to be interested in animals; more often, they have to learn which ones flee when approached and which ones bite.[15] Our interest in other species is a characteristic of our own species. To say that it is part of our DNA is probably more than just an overworked metaphor. How that interest, that affiliative predisposition, is implemented is a cultural phenomenon, varying in intensity and form from population to population, just as tool-making is something fundamentally human, but is manifested in the diversity of our past and present material culture.

Figure 55. As squirrels have learned to take food put out for garden birds, the technology of birdfeeders has adapted—an example of cultural coevolution. (Source: author.)

It is tempting, but naively reductive, to seek to explain every distinctive attribute of human behavior in evolutionary terms. However, something of that nature must underlie the generality of zoophilia throughout our species. What is its purpose? What is or was its adaptive value? Here we have to retreat into speculation. While we can assess some attributes, such as skin pigmentation or blood groups, we cannot assess the function or phylogeny of zoophilia by examining situations in which it is absent or takes a wholly different state. All we can say is that it is a characteristic that humans acquired at some point in their evolution and that we have retained, albeit with population-specific adaptation. A plausible explanation would be to suppose that it was advantageous to our species to take an interest in others in order to assess their potential as resource or threat. Such a behavioral trait would have been valuable when *Homo sapiens* was extending its global range during the Late Pleistocene, encountering new environments and faunas, and coping with the climatic vicissitudes of the Last Interglacial and succeeding cold stage. Those circumstances could have imposed quite a strong selection pressure favoring the evolution of a hard-wired predisposition to take an interest in other species, familiar or not, the better to learn their potential utility or hazard. To take the speculation one step further, perhaps this was one of the behavioral differences between our species and Neanderthals, one that gave us the edge at a time of rapid environmental change. Finds of cut-marked bird bones from Neanderthal levels in Fumane Cave, Italy, have recently been proposed to indicate the collection of feathers, "an activity linked to the symbolic sphere."[16] From what we have already seen, it should be no surprise that alpine choughs *Pyrrhocorax graculus* were particularly abundant among the bird bones. If we accept that the cut marks genuinely represent some systematic dissection of the wings of several species of birds, that is still some distance from prime facie evidence for feather-based symbolic activity, and hence from inferring a modern form of zoophilia in Neanderthals. Pending better evidence, we can safely say that zoophilia is a characteristic of *Homo sapiens*, but that its existence or implementation in other hominin species remains a matter of interesting speculation.

If genuine indifference to other species is the exception rather than the rule, species that imposed themselves on our attention by moving in as unprompted neighbors or even as domiciliary lodgers must have provoked a significant cultural response. Alongside the utilitarian, essential prey or livestock species, the commensal animals became the species of the everyday scene. The tendency for people, especially adolescent girls, to exhibit a nurturing response towards the young of other species is well-known. A number of evolutionary arguments have been put forward to account for this, not least the idea that girls are demonstrating their potential as mothers by exhibiting extraspecific nurturant behavior.[17] People in general, and girls in particular, seem to be drawn to animals that exhibit similar visual cues to human babies: namely, a large head relative to body size, and large, forward-facing eyes. A warm body, especially with the addition of fur or fluff, also seems to be an important attractive trait. This "cute" response may be one particular aspect of a more general cognitive attribute of human beings—namely, the capacity to "think like an animal," which Steven Mithen has argued developed as a part of becoming modern humans.[18] In the ethnographic literature, there are copious examples of hunter-gatherer people "adopting" and caring for the young of species that they also hunt as prey. Bradshaw and Paul[19] draw attention to the interesting case of the Guaja people of the Amazon, who adopt baby monkeys, which look not unlike baby humans, to the extent of feeding and even suckling them. The Guaja also have dogs, but the dogs are treated relatively badly: their main source of food appears to be scraps dropped by the pet monkeys. The Guaja example is a formal, not a relational, analogy. It tells us nothing specific about earlier peoples, though it is a useful reminder of just how complex and irrational human association with other species can be, and hence the difficulty of arriving at a convincing evolutionary explanation.

A less overtly Darwinian way of viewing our affinity for other animals is to propose that human beings have a deeply ingrained, almost spiritual bond with other living organisms of all kinds—the *biophilia hypothesis* proposed by the great zoologist Edward O. Wilson in the 1980s.[20] It seems to be the fate of innovative big ideas that they are used as a framework on which to hang a far wider range of inferences and applications than had originally been intended: Marxism, Christianity, and the Gaia Hypothesis have all suffered in this way. Biophilia, too, has become a prism through which a remarkable diversity of attitudes towards conservation and environmentalism have been projected. For example, in an influential essay on the subject, American ecologist Michael Soulé observed that "If biophilia is destined to become a powerful force for conservation, then it must become a religion-like movement."[21] Wilson's eponymous book is a quite short compilation of linked essays that serve to illustrate what the author sees as humanity's "innate tendency to focus on life and life-like processes" by instructive anecdotes and recollections from Wilson's own remarkable career. There is no concise executive summary, and no specific take-home message beyond the uncontentious one that life on Earth is richly diverse and wonderful, and that to study and to cherish that diversity is good for the human . . . what? Soul?

Biophilia, as proposed by Wilson, is ethical rather than utilitarian, driven by deep-seated psychological traits that humans acquired at an earlier time in our evolution and that we now manifest in many aspects of our behavior. Our affinity for other animals, which I have proposed to have been an important element in facilitating the development of commensal relationships, could be seen as a facet of Wilsonian biophilia. To explore this a little further, Stephen Kellert proposes nine different dimensions to biophilia (utilitarian, symbolic, aesthetic, and so on), of which the naturalistic tendency (satisfaction derived from direct contact with nature) may be the most relevant to our consideration of commensal species.[22] The commensal species were not chosen by us: they chose us, so they are unlikely to be of interest primarily on aesthetic grounds. Nor are they necessarily useful, although we have subsequently made use of some of them, reflecting Kellert's utilitarian dimension. Today, to the extent that we value at least some commensal animals at all, it is because they fulfill our apparent need to have direct contact with the natural world. This can hardly have been so in the prehistoric past, when human detachment from "nature" was presumably less marked, though perhaps there was a positive cultural response to species that showed a reduced flight response without posing a direct threat to people. This begs the question of whether prehistoric peoples made any distinction between their own domestic world and the natural world around it. If they did not, then the commensal animals that we might regard as transgressing the wild/domestic boundary would have had no such significance.

Much has been written from an ethnohistorical perspective to argue that native, aboriginal, or First Nations people take a more holistic view of the living and nonliving world, and of themselves as a fundamental part of that world. Some take that further, attributing to such people a degree of wisdom about the "natural world" that we industrialized Westerners lack and must relearn. Richard Nelson typifies this attitude, writing *inter alia* that "much of the human lifeway over the past several million years lies beyond the grasp of urbanized Western peoples,"[23] and that "it's essential that we learn from traditional societies, especially those in which most people experience daily and intimate contact with the land . . . In such communities we find knowledge similar to that achieved by our own scientific disciplines . . . insights founded on a wisdom that we had long forgotten and are now beginning to rediscover."[24] I would argue almost the opposite: that urbanized Westerners probably understand more about the evolution of human lifeways over the past several million years than did the Native American peoples studied by Nelson and other ethnographers. Furthermore, the knowledge of such peoples, though deep and detailed, is highly context-specific, and so often founded in myth and spirituality that any comparison with scientific disciplines would be a serious category error. I do not deny for one moment that people who engage in hunting, fishing, gathering, and trapping for subsistence develop a deep and detailed knowledge of the world in which they live, and often appreciate its ecological subtleties to a remarkable degree. However, this knowledge is, as I have said, highly context-specific, and so unlikely to yield a paradigm of global applicability. More specifically, it would be unwise to suppose that attitudes to other species manifested even by First Nations people today bear any resemblance to attitudes shown by prehistoric peoples, other than the generalized cross-cultural interest in other species to which I have repeatedly referred, and which is seen just as clearly in industrialized Western cultures.

One of the challenges faced by proponents of the biophilia hypothesis is that it has become entangled with this "native wisdom" tendency, thus implanting situationalist narratives onto what is supposed to be a generalized model. Another is that Wilson originally used the term "hypothesis," giving his critics an easy target. Yannick Joye and Andreas de Block, for example, make the point that any hypothesis has, by definition, to be open to falsification, and that the often imprecise and ambiguous terminology used by advocates of Wilsonian biophilia prevents any explicit falsification.[25] Others have been critical of the idea that biophilia could be harnessed to drive biodiversity conservation efforts,[26] some even arguing the opposite on the grounds that most human interaction with other animals over the last few million years has consisted of sneaking up on them and killing them.[27] Harold Herzog gives a more measured and thoughtful view.[28] Biophilia, he argues, is not a tenable generalization because of domain specificity. Depending on the circumstances in which they lived, our ancestors would have needed to develop a range of mental modules to best accommodate to the species around them. On the plains of Pleistocene Africa, for example, animals would have been prey, or foes, or helpers such as wild canids that cleaned up food debris around camp and gave warning of large predators. Herzog offers an evolutionary-psychology model based on positive or negative biotaxis, an innate tendency to move towards or away from other species. A capacity for biotaxis could be inherent, a hardwired part of being human, but the implementation (approach cats, avoid snakes) would have a learned, cultural component. What Herzog and other authors do not explore in any depth is the likely biotaxic, or biophilic, response to species that actively intrude upon our living space, not waiting to be approached or avoided, but making the first move to their own behavioral agenda.

When another species is regularly present in our living space, the human response will be to take note and to act—either positively, perhaps by supplemental feeding, or negatively, by attempting extirpation. The latter reaction raises the question of how and why some species become regarded as "vermin" and therefore the target of an often lethal reaction. The obvious answer is that vermin species are those that eat our food to an unacceptable extent and/or convey disease. We have seen that the alleged role of some commensal animals as disease vectors has been the rationale behind extermination campaigns. The association of rats with plague and, latterly, with Weil's disease, or of pigeons with cryptosporidiosis has been firmly entrenched in the public mind in Western cultures, albeit probably overstated. But would the potential disease risk have been a factor in the behavior of people who did not have a "germ model" of disease transmission? The use of mice in Ancient Egyptian medicine shows that even a culture that had a detailed and sophisticated knowledge of the structure and functioning of the human body could show apparent nonchalance regarding the zoonotic disease risk of ingesting uncooked, whole rodents. It seems unlikely that modern concerns regarding disease and "dirt" would have figured in past reactions to commensal animals, as these concerns are based on a knowledge of disease mechanisms and vectors that essentially derives from Western Europe over the last two centuries.

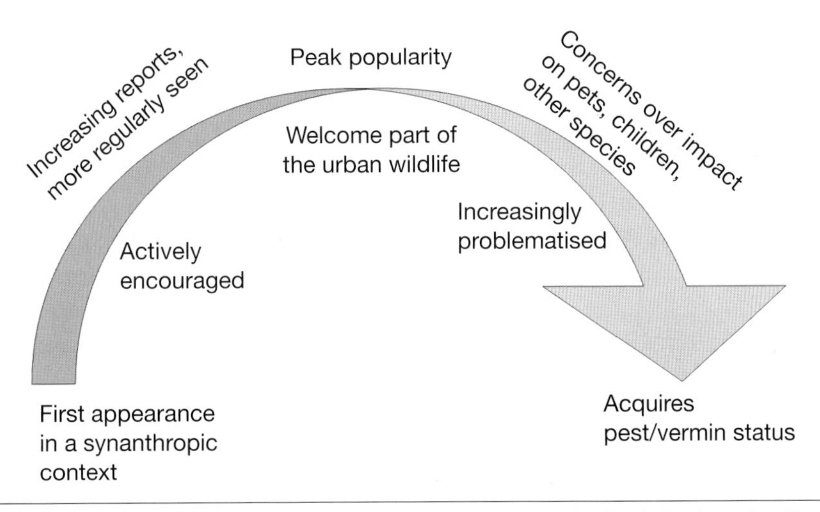

Figure 56. The "arc of acceptance," which seems to apply to so many commensal animals that have the effrontery to adapt successfully.

To what extent, then, did prehistoric peoples react against food-stealing by commensal animals? The point has already been made that loss of crops to "pest" species may be more tolerable to subsistence farmers than to those for whom any crop surplus to their own subsistence needs can be traded. If mice eat 10 percent of your stored grain, that may be a disaster if the stored resource is barely sufficient in the first place, but tolerable if a good harvest means that there is plenty in store. And if a cat, mongoose, or civet adopts your home as a good source of mice and so reduces that loss to less than 5 percent, so much the better. The point is that reactions to commensal animals may have been situation-specific, even varying year to year, and the presence of one such species may have affected the response to another. The simple categorization of some species as "vermin" may be a largely modern, Western concept, one which has no place in any longer-term consideration of the place of commensal animals in our world. The abundance of house-mouse bones at the famous Turkish site of Çatal Hüyük could lead us to think that the Neolithic inhabitants "had a mouse problem." We might go further, noting the substantial accumulations of occupation debris between and around domestic buildings,[29] and concluding that the Neolithites lived in utter squalor, surrounded by garbage and therefore overrun by mice. However, it is entirely possible that the mice and the garbage were tolerated, were even seen as an appropriate part of the human living environment. An ethnographer from a quite different culture and from several millennia hence might wonder why I not only tolerate large numbers of small birds in the vicinity of my home, but actually put out food to attract them. Did these Brits know nothing of zoonotic microorganisms and feather lice?

The acceptability of, and therefore attitude towards, commensal animals is highly culture-specific, perhaps even highly individualistic. Coyotes have already been mentioned as a species that stretches that acceptability to its limit, and these canids are the subject of one of the most detailed and wide-ranging discussions of the topic.[30] Judith Webster explores the origins and outlook of the city of Vancouver's "Co-Existing with Coyotes" program, of which she is clearly not enamored. Analyses of coyote scats and numerous finds of cat collars at coyote den sites on the edge of towns show that coyotes regularly prey on cats. There seems to be little room for disagreement on that point, although statistics can be marshaled both to talk up and to minimize the scale of the predation. Similarly, cats eat small birds, and statistics are regularly published that purport to show that this has a serious, or an insignificant, impact on small-bird mortality. For promoters of urban coyotes, Webster angrily notes, the consumption of cats is to the credit of the coyotes, because they thereby promote the survival of birds. This side effect of "mesopredator release" was explicitly identified as a consequence of reduced coyote numbers in the San Diego region, in a study that also drew attention to the high density of cats maintained by feeding "subsidies" from their human keepers (or domestic staff).[31] There are much larger questions here of who decides whether a particular commensal species is vermin or welcome neighbor, and of the tactics that are used by different sides of the debate to convince others. "Deep-Green" environmentalists tend to demonize cats as recreational bird-murderers (though one suspects that some of them dislike humans equally strongly), while pro-pet groups point out that a well-fed cat will have little reason to hunt. Neither of those polarized extremes is susceptible to persuasion; what is interesting here is the way that such debates tend to focus on quite simplistic models of human/animal interactions in urbanized areas, to the detriment of a richer, more nuanced understanding.

Questions of terminology and classification were raised at the beginning of this book, and this is the place to revisit them. Can we develop a terminology to describe the spectrum of relationships that people develop with their animal neighbors, one that is precise enough to be useful, while sufficiently flexible to be applicable?

Taking animals as a whole, we conventionally divide them into wild and domestic animals, terms that are usually applied at species level ("Tigers are wild animals, sheep are domestic animals"). This is clearly inadequate for any subtle analysis of human-animal affiliations. At a first level of classification, we need to differentiate those animals that are owned, captive-bred, and fed as subsistence resources: the cattle, sheep, pigs, horses of the farmyard, and their ecological replacements in other biomes. The term *livestock* commends itself: these are domesticated animals, but to be differentiated from those domesticated animals that share our homes and lives as *companion animals*. This differentiation is not necessarily at species level. Although we might cavil at referring to dogs as livestock, that is the most plausible interpretation of archaeological remains at the site of An Son in Vietnam, where dogs appear to have been captive-bred, fed, and then butchered for food.[32] Dogs can be livestock, and lambs can be pets: the relationship is the defining criterion, not the species.

Pets and livestock aside, the "wild" animals divide into two categories: Those that willingly live in close association with people, utilizing the modified or constructed environment of human habitations for living space and food, are the *synanthropes* or *synanthropic animals*—again, defined by population, not by species. It follows that there are others, the most obviously "wild" animals, which actively avoid places where environmental perturbation or direct disturbance by people is likely: the term *antanthropes* (*antanthropic*) serves for them. Among the synanthropes, we find the species that are the subject of this book, though a further level of differentiation is necessary. *Commensal* animals, as discussed in the Introduction, derive food benefit from their association with us. Some animal populations live closely alongside people, but predominantly for the benefit of living space and shelter, rather than food. Several of the bird species discussed in chapter 6 fall into this category, such as the larger gulls that use buildings as protected nesting space but seem to gain little feeding benefit from doing so, or the starlings that roost on city-center buildings by night and disperse into the surrounding landscape to feed by day. A good mammalian example of this group is the pipistrelle bat *Pipistrellus pipistrellus*. Pips form large breeding and roosting colonies within houses and churches, using the roof spaces as other bat species use hollow trees and caves. They disperse at night to feed, often around the same buildings, but not acquiring food directly from human activities except insofar as our streetlights attract and confuse plump, delicious moths. This is so obviously a different adaptation to that of the larder-raiding mice as to need a different term: one could hardly describe pips as commensal. The term *edificarian* (derived from "edifice," i.e., a building) is applied to animals such as geckos that live much of their lives on the constructed surfaces of buildings, and that term can usefully be transferred to animals that inhabit the exteriors and sometimes interiors of our buildings and other constructions for the benefit of living space rather than food. Thus, synanthropes may be commensal or edificarian.

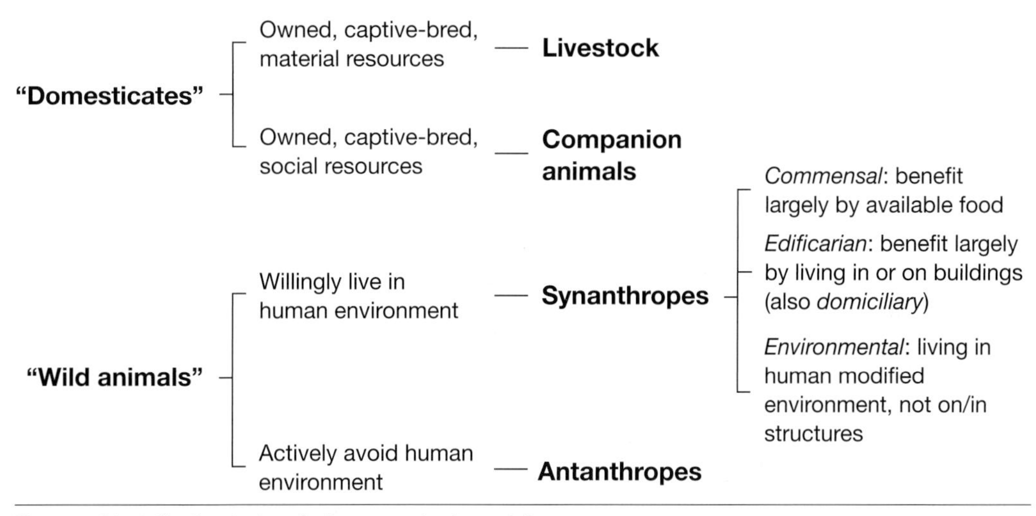

Figure 57. A tentative terminology for human-animal associations.

There are other necessary qualifying terms, too. Synanthropic animals can be divided between those that live within or directly upon people's domestic structures (*domiciliary*) and those that live within the modified human environment (*environmental*). Edificarian animals are often domiciliary, though not necessarily so, and commensal animals can certainly be either. Urban foxes, for example, are environmental commensals. Another distinction has already been drawn between animals that have passively entered a synanthropic role because all or part of their preexisting range has been settled in or modified by people (*adapters*), and those that have actively extended their preexisting range in order to exploit opportunities created by human settlement or perturbation (*adopters*). To stay with foxes, the evidence would seem to suggest that they are adopters, actively and opportunistically moving into the human environment.

Chapter 3 asked for how long all of this association has been going on. Taking the terminology developed above, the first commensal animals to take advantage of people in the Pleistocene were likely adopters rather than adapters. The density of human activity at that time is unlikely to have necessitated an adaptive response, though it would have offered opportunities. In the absence of buildings or other substantial construction, the first animals to associate with people were probably environmental commensals. Edificarian and domiciliary adaptations could not have arisen before people were living in built structures, and hence not before the Natufian and early Neolithic. The association of rats and foxes with Natufian settlements in caves and rock shelters is somewhat open to debate, but was probably environmental commensalism—by then a well-established adaptation on the part of some animal populations. Domiciliary commensalism probably began with mice moving into Neolithic structures: the abundance and distribution of mouse remains at early sites such as Çatal Hüyük certainly indicates that they were living within, not only around, the clustered dwellings. Whether cats became domiciliary from the outset, or had a long period of time as environmental commensals, is not clear from the current archaeological record.

The last element of the repertoire to arise was probably the adapter, rather than adopter, strategy. Presumably that arose only as the human population attained a sufficient density in at least some places to make adapt-and-stay a more successful adaptation than move-away. As patches of unperturbed habitat became more fragmented and smaller in extent, so adaptation to a synanthropic life would have become more and more favorable. That point may have been reached in later prehistory in some parts of the world (the Levant, coastal Peru, along the great rivers of China), only relatively recently in others (East Africa, mainland Southeast Asia, California), and perhaps not at all in some places (Boreal Europe, Central Asia?). There is an intriguing archaeological study to be undertaken in those late-adapter regions to test the hypothesis that, for example, the settler towns that Imperial Russia pushed into Siberia and the Boreal regions would have acquired a probably domiciliary commensal fauna recognizable as part of the colonizing "package" and only the most opportunistic adopters from the local fauna. The relatively impoverished commensal faunas that we see in Roman towns in Britain (chapter 3) may be analogous. In previous chapters, we have considered the possibility that competition between would-be commensals may have inhibited or excluded some species, such as house mice being excluded by *Apodemus* mice in later prehistoric Europe, or urban foxes being excluded by feral urban cats in medieval towns. Further research into patterns of competitive exclusion and character displacement in modern commensal faunas would certainly help us to understand the origin and development of the various affiliations that we see in the archaeological record.

By taking the long view of commensalism, it is easier to see it as a cultural coevolution, as millennia of learning to live together, rather than as some epiphenomenon of the way that we live today. And by stressing that the adaptation was and is at population, rather than species, level, it is easier to understand that the current situation does not fully define or constrain what may have happened in the past, nor what may happen in the future.

8

Planning for the Future

ALTHOUGH THIS BOOK IS CONCERNED WITH THE PAST AND PRESENT OF COMMENSAL animals, the present is only a very thin interface between past and future, so what of the future? The long story of our animal neighbors makes it clear that some will cope with whatever changes we make to our world. As with so much in animal conservation, the question is not so much whether we want a commensal fauna as what sort of commensal fauna do we want and what will we do with it?

Does it matter whether or not the human settlements of the future are shared with gate-crashing neighbors? I would argue that it does. As we have seen, opinions are diverse and conflicting with regard to the Biophilia Hypothesis and the extent to which it describes something fundamental about our species. Nonetheless, there is a body of evidence that makes it clear that interaction with other species is something that people will seek in some form, and from which we benefit in a number of ways. As we become more and more an urban species ourselves, the animals that adapt to live with and among us are the ones that will fill this beneficial role. Of course, companion animals will continue to be important. In his strange and insightful novel *Do Androids Dream of Electric Sheep?*,[1] Philip K. Dick envisaged a world in which highly realistic robotic animals fill the role of companion animals, and the ownership of a real animal becomes the stuff of prestige and the black market. What Dick hit upon was the altogether believable notion that in a densely urbanized world in which most people have no contact with real animals, people would have to invent a replacement, as much like the real thing as possible. However, even the most realistic and subtly engineered of these artificial animals fails to fulfill the need of at least some of the novel's characters for direct contact with real, live animals. The same question, whether "technological nature" can replace the real thing, is explored more academically, though less entertainingly, in the psychological-research literature.[2] What Dick does not explore in the novel is the consequences for human health and behavior of having *only* "owned" companion animals in the everyday experience, and no animals that come and go of their own accord, "wild" yet sharing the human living space. Perhaps more to the point, these animals are a tangible part of "nature" with which we can interact, rather than part of the "home and family," in which role we generally see companion animals.

Using the terminology developed for chapter 2, there is both anecdotal and scientific evidence for the therapeutic benefit of close and regular contact with animals. I suspect that at least some research in this area suffers from the "common sense" presupposition that contact with animals is good for you, so only the most contradictory evidence would attract notice. That said, there are well-controlled studies to show, for example, that in a program of outdoor activities conducted with and without the involvement of dolphins, patients suffering with mild to moderate depression showed an improvement of symptoms, but more so when the dolphins were part of the therapy.[3] The therapeutic value of allowing companion animals in hospitals has been known for several decades,[4] as has the value of incorporating animals into psychotherapy treatments.[5] If we place those animals in a "natural" context, we come close to one of the predictions (or assertions?) of the Biophilia Hypothesis, as argued by, for example, Eleonora Gullone.[6] Without taking sides in this debate, it is clear that if contact with animals is beneficial, and if regular experience of "nature" is beneficial, then any manifestation of the "wild" that manages to persist within our highly modified living space is of more than just curiosity value, and is likely to have therapeutic value for at least some people in any human population.

Figure 58. The therapeutic role of animals such as Kozmo the miniature pony is widely understood and appreciated. (Source: Cassandra Wilson, WeeWhinnies Therapeutic Minis, with permission.)

Perhaps, too, our commensal neighbors should be valued for their educational role. The increasing isolation of city-raised children from the ecosystems that support and feed them has been well rehearsed in the literature.[7] Quite apart from the psychological consequences for the next generation, there is a significant challenge in trying to maintain understanding of, and concern for, the conservation of biodiversity and wildlife in young people for whom "nature" is images on a screen, and not a reality to be heard, seen, smelled, and touched. That, too, links to debates within evolutionary psychology over the role of animals in moderating behavior towards other people, and the notion that a child that behaves cruelly or dismissively towards other animals may be more likely to develop psychopathic behavior towards other people.[8] Even if we regard that as speculative, the argument remains that we need the next generation and the one after that to value the diversity and continuation of life on Earth, and most of the people that grow up in those generations will do so in cities. The animals that they see, apart from companion animals in their homes and guard dogs on the streets, will be their commensal neighbors. For the sake of the education of the next generations, then, our settlements need a diverse and abundant fauna.

Beatley's concept of *biophilic urbanism* is a step in that direction.[9] Granted, what Beatley argues for is the incorporation of "natural and biophilic elements . . . into everything we design and build,"[10] making the more general point that patches of green landscape can be placed in buildings, neighborhoods, and cities, to facilitate health and educational benefits. That premise is, it seems to me, unarguable. Even if one is skeptical about the therapeutic benefits of green patches, the measurable, practical benefits of mediating pollutants and moderating surface-water runoff ought to make the case. From all that we have seen, it is clear that commensal animals will colonize even modestly biophilic settlements. The interesting challenge will lie in fine-tuning those green spaces and buildings to allow the greatest diversity of species to establish and sustain populations. In particular, we have seen that many of the animals concerned are literally commensal—they derive food from our wastes and other activities. The truly biophilic town would have plenty of accessible garbage for animals to scavenge, and to sustain the invertebrates that would attract still other species, but that is unlikely to meet with general public approval, let alone civic-planning consent.

That brings us to another point. Adapting to biophilic urbanism will not simply be a matter of designing the right buildings and open spaces, but will require some cultural adaptations on our part. Beatley writes[11] that obstacles to green cities in the United States are mainly aesthetic, with "nature" being seen as untidy. Part of the challenge will be to shift public opinion, or at least the planners' opinion, to be less tidy, less precious about untidy green corners, and less anxious about scavenger access to garbage when that access can be managed to be off the street and not too intrusive. We may, too, have to be less anxious about the potential danger inherent in sharing our space with some of the larger animals. No feliphile wants to see their companion eaten by a coyote,[12] but even these large and potentially dangerous animals could be accommodated if coyote populations can be prevented from becoming too well habituated to people. Populations of people and of coyotes need to find a compromise in which the coyotes gain access to garbage, windfall apples, and rodents, yet learn avoidance of humans, so as to minimize direct conflict. Once again, this is a matter of cultural coevolution, enacted at the population level, not generalized across people and coyotes as a whole. Scrupulously clean, tidy, safe towns and cities that we share with few or no other animals apart from our pets would not be healthy places to live.

Figure 59. Feeding the pigeons in Krakow, Poland, is an example of positive engagement with urban commensal wildlife. (Source: Christopher Walker, Fotopedia, with permission.)

There is a more academic case, too. In a world in which half of the human population live in cities, and most of the other half in towns or large villages, it could be argued that the future of animal conservation lies in the human modified environment, and therefore with the species that have best adapted to that environment. That is not to say that we should abandon attempts to retain large patches of major biomes with the minimum of human impact, only that we should recognize the reality of the situation and not neglect the opportunity that the human environment and its commensal fauna represents. After all, this is not something new: we have had commensal neighbors at least as long as we have had agriculture, and arguably much longer. Commensal animals are the ones that have best adapted to our living space; as that living space extends to more of Earth's land surface, it behooves us to understand how to facilitate its adoption by more species. Some of those other species will be strongly antanthropic, and there may be little that we can do in conservation terms other than to declare large refuge areas and ensure that those areas are maintained. The conflict for living space that has arisen in several parts of India between people and tigers, for example, is never going to be resolved by even the most brilliant of biophilic urban planning. Sharing the streets with coyotes is difficult enough. For many other animal populations, however, there may be ways of ameliorating the impact of human contact, in some cases even by facilitating a route to commensalism that allows a particular population to persist, albeit behaviorally altered: to become adapters. That amelioration may be through development planning or architecture, or it may include an element of behavioral adaptation on our part. The successful commensal adaptation of some monkey species in India and adjacent countries, for example, is at least in part a consequence of human adaptation deep in the past, developing ethical and religious codes that allowed space for pilfering macaques and langurs. In order to facilitate such an adaptation on the part of challenged animal populations, we need to understand how commensalism works for those successful species, the better to encourage and facilitate it in others. The point has already been made that much of the detailed scientific literature on commensal animals has its basis in a perceived need to "control" (for which read reduce, eradicate, extirpate) those animals. Nonetheless, there is much in that literature to show that the biology of commensalism is a rich and significant topic for research. The better we understand that topic, the more we may be able to utilize that understanding to the benefit of animal populations.

That said, a degree of humility will be required. The point about our commensal neighbors is that they have moved in without invitation. Although we can, and should, take steps to allow space and food for a diverse range of animals within our living space, we will have to accept that in the end, they will decide whether or not to come. The archaeological and historical records show that opportunities will not always be taken. The surprising dearth of foxes in medieval towns in England may, as suggested in chapter 4, be because of competitive exclusion by feral cats, but we cannot rule out the possibility of other, quite subtle factors. The relatively impoverished commensal faunas of Roman towns in England, with little more than the domiciliary rodents, probably tell us something about the intensity of use of space and an intolerance of accessible garbage, though that remains an untested postulation. Conversely, the adoption of suburban habitats by goshawks *Accipiter gentilis* in modern Germany is quite counterintuitive.[13] We can provide the opportunities, and undertake the research to ensure that those opportunities are fine-tuned, but the other animals involved will respond as suits them, not as suits us.

Further research into commensal animals could usefully move beyond individual species to explore the community as a whole. In doing so, it becomes possible to separate the topic from any geographical specificity, to consider the commensal strategy per se rather than particular species. Defined by a largely donor-controlled food web, this animal guild has highly characteristic community dynamics, and distinctive patterns of competition, exclusion, and predation. Within the commensal guild there will be degrees of facilitation as well. There is a need, too, for further research into people's attitudes towards commensal animals—again, not at the species level, but as a whole. Much of the literature on this contentious matter tends to focus on this or that species—pigeons, foxes, monkeys—and, as I have said, to be dominated by questions of control and extirpation. What about the commensal community as a whole? How is it regarded by scientists and by the public? Are commensal animals "wildlife," a part of the natural world that happens to be interdigitated with our living environment, or are they regarded differently? TV nature documentaries in Britain have begun to include features on, for example and most often, urban foxes alongside the more usual wildlife offerings. This may indicate a broadening of perspectives among those who commission and direct such programs, though the ready availability of viewers' photographs and video clips may also be an encouragement. In the Upper Paleolithic, the first commensal adopters may have become the animals that most people most often *saw*: today, they have become the animals that people most often *video*. This also means, of course, that the commensal animals become the ones most readily incorporated into "citizen science" projects,[14] though it would be a missed opportunity if a rift emerged between amateur biological studies of foxes and pigeons, and professional studies of less-familiar neighborly animals. We need a good integration of anecdotal and scientific studies of all animals.

There are still questions to be addressed regarding the transition from neighbor to vermin, and if we wish to incorporate animals in our new, "greener" towns and cities, this transition is particularly important. I have argued in earlier chapters that the association of some animals with a disease risk has sometimes been a post hoc justification for a change of attitude that was already in place. There is scope here for more detailed historical studies. The association of, for example, rats with disease has a genuine basis, as rats are vectors for a number of significant zoonoses. But that is scientific knowledge, known and understood by some of the general public, but unlikely to be clearly known by everyone who, if asked, would regard rats as vermin. There may be a more general association with untidiness, squalor, and garbage that factors quite directly into both culturally driven concepts of acceptability, and deeper, more hard-wired risk-avoidance behaviors. That is a quasi-evolutionary explanation. Colin Jerolmack[15] takes a different slant, proposing that modernity requires a firm boundary between nature and culture, and so any animals that transgress that boundary become a problem: pigeons are the antithesis of the orderly and sanitized ideal metropolis. I think that explanation stands in want of evidence. Do people other than architects and planners actually *want* to live in an orderly and sanitized metropolis? People like animals—not just in their place in "nature," but in and around our homes as well. For some reason, we take particular exception to certain of those animals, but for reasons that are clearly more subtle and complex than simply their transgression of "our" space, and that perhaps have more to do with the behavior of those animals and the acceptability of that behavior were we to encounter it in ourselves. Most people would be at least mildly disgusted at the idea of rummaging through supermarket garbage bins in search of edible food, even though that food may be perfectly edible and, best of all, free. Maybe it is fanciful to hope that the trend towards more recycling will help us to revise attitudes towards animals that eat food and other wastes that we have rejected! Fanciful or not, the concept of "vermin," and the factors that lead a species to be regarded in those terms need to be clearly understood and managed by would-be urban greeners.

Understanding the past is important here. Biotic homogenization means that we have, perhaps, a less diverse range of commensal faunas than may have occurred in the past. The record of past commensal faunas can show cultural and ecological adaptations on the part of other species that may not be apparent today for want of the right circumstances, but that may, therefore, be possible in the future. In Britain, we have successfully reintroduced the red kite *Milvus milvus* to rural areas from which it had been previously extirpated, and an uplifting sight they are, too. But the archaeological record clearly shows that kites were a regular part of the urban fauna in medieval times: could that be the case again? And if we could reestablish conditions that would bring kites back into our towns, would we necessarily want them? Direct observational research, whether investigating animals or people's attitudes to animals, can only take a perspective of a few years' duration, albeit well controlled and in great detail. At the decades to centuries scale, the *longue durée* of historians, we need other sources. That requires historians and archaeologists to take an interest in species that were neither prey nor livestock nor our chosen companions, but that were, nonetheless, our neighbors.

Notes

INTRODUCTION

1. See Juliet Clutton-Brock 1999 and 2012.
2. See Honeycutt 2010 for a useful summary of current genetic evidence.
3. Shipman 2011.
4. For example, Anderson 1989 details the human contribution to the extinction of moas in New Zealand. Sodikoff 2009 provides a series of thoughtful essays on different aspects of extinction.
5. For example, in Thompson 1989.
6. Boucher et al. 1982 give an excellent overview of this topic. See also De Mazancourt et al. 2004; Dickman 1992.
7. García et al. 2010.
8. Boucher et al. 1982.
9. Alberti et al. 2008, 150.
10. McKinney 2006.
11. Johnston 2001.
12. McKinney 2006.
13. O'Connor 2010a.
14. O'Connor 1997; Russell 2002; Zeder 2006; Driscoll et al. 2009.
15. Gentry et al. 2004.
16. O'Connor 1997; Budiansky 1992; see also discussion of domestication in Cassidy and Mullin 2007.
17. Driscoll et al. 2009 and Trut et al. 2009 give somewhat differing views on the relationship between "tame" and "domestic." See also Réale et al. 2007 on the heritability and evolutionary significance of "tameness."
18. Kajiura et al. 2009.
19. Andrezjewski et al. 1978; Luniak 2004.
20. Smith and Kenward 2011 discuss this adaptation in the context of Roman Britain; see also Ezequiel et al. 2001 for a very different context.
21. Bleed 2006; O'Connor 2000a.
22. Bonanni et al. 2007.
23. For example, urban gulls; Indykiewciz 2005.
24. Banks et al. 2010.
25. For example, the annual "Garden Bird Survey" run in the UK, or widespread sales of "Garden Bird" food. Advertising for the latter and the iconography of both make clear that the category includes small, attractive, mainly seed- or invertebrate-eating species, not hawks or nest-robbing crows.
26. Ingold, in Manning and Serpell 1994.
27. Conard 2003.
28. Arnold et al. 2006; Miller 2008; Marzluff et al. 2008.
29. This is discussed at greater length in chapter 8 of this volume.

CHAPTER 1. THE HUMAN ENVIRONMENT

1. Crutzen 2002; Steffen et al. 2007.
2. Crutzen 2002.
3. Ruddiman 2007; Ellis 2011.
4. Wilson 2007.
5. Jones et al. 1994; Wright and Jones 2006.
6. Arnold et al. 2006; Gilbert 1991; Shochat et al. 2006.
7. O'Connor 2011.
8. This is discussed further in O'Connor and Evans 2005.
9. Niemelä 1999; Rebele 1994.
10. O'Connor 2010b.
11. Roxburgh et al. 2004; Svensson et al. 2007; Graham and Duda 2011.
12. Irwin et al. 2006; Ruiz et al. 2007.
13. Smout 2003; McKinney 2006; Long 1981 and 2003; Flannery and White 1991.
14. Sykes 2004 and 2010.
15. Jerolmack 2008. Central to this paper is the questionable assertion that "modernity posits a firm boundary between nature and culture."
16. Biotic homogenization—McKinney 2006.
17. Rathje and Murphy 1992; Rathje et al. 1992.
18. Evans 2010.
19. O'Connor 2000b further explores this topic.
20. For example, the stinging nettle *Urtica dioica*; Taylor 2009.
21. Hutchinson 1978; O'Connor and Evans 2005, 28–29.
22. Polis and Strong 1996 give an excellent discussion of food-web dynamics.
23. Fahrig and Rytwinski 2009; Benítez-López et al. 2010; Osłowksi 2008; Kociolek et al. 2010.
24. This contentious issue is usefully summarized by Douglas and Sadler 2011.
25. Arnold et al. 2006.
26. Originally used by Le Corbusier (Charles-Edouard Jeanneret) in his 1924 *Vers un architecture* (Paris: G. Crès), and reiterated ever since.
27. Rubble sites; the classic studies of urban rubble sites are those undertaken by Owen Gilbert 1991 in Sheffield, UK. See also Sukopp 2008.
28. Pigeons as a problem—a recent, concise account is given by Haag-Wackernagel and Geigenfand 2008.
29. A useful summary is given by Lundholm and Richardson 2010.
30. Soldatini et al. 2008.

CHAPTER 2. SOURCES OF EVIDENCE

1. This huge topic is summarized by Curthoys and Docker 2010; Isacoff 2005 discusses a specific case study.
2. A fine example of combining sources is given by Boivin et al. 2009.
3. Potts 1967.
4. Clergeau and Quenot 2007.
5. McClure 2011.

6. Though as he is now developing an interest in "greening" cities, all is not lost.
7. Gehrt 2007; Gehrt et al. 2011.
8. Murton et al. 1972b.
9. Haag-Wackernagel 2006; Haag-Wackernagel and Moch 2004.
10. Jacquin et al. 2010.
11. Taphonomy is dealt with at length and in detail in sources such as Lyman 1994; Martin 1999; and Allison and Bottjer 2011.
12. Discussed by Buczacki 2002, 477.
13. White and Greenoak 1988.
14. This process of replacement is discussed in more detail in chapter 5.
15. See Berry 1987. James Fisher is quoted thus in Thornton 1997, xi, though without a definitive source. I have taken this quote at face value because the style and the sentiments are so recognizably Fisher's.
16. Bell 1837.
17. Bell 1837, 78.
18. Bell 1837, 182–85.
19. Buczacki 2002, 296.
20. Discussed and given a wider context by Sim 2005, 2.
21. For a general introduction to the study of animal bones from archaeological excavations, see Reitz and Wing 2006; O'Connor 2000a.
22. Discussed in O'Connor 2000a, chap. 3.
23. For example, Horsburgh 2008.
24. Buckley et al. 2010.
25. For example, see Bar-Oz and Tepper 2010.
26. Reinhard and Bryant 1992 give a useful overview.
27. Sardella and Fugassa 2009.
28. Pruvost et al. 2011.
29. Kimura et al. 2010 give a nice case study on donkeys; Brown and Brown 2011 give an overview of the field and more besides.
30. Yu and Peng 2002.

CHAPTER 3. THE ARCHAEOLOGY OF COMMENSALISM

1. Hamilakis 2001.
2. For example, the heavy-metal fallout from prehistoric mining; Mighall et al. 2009.
3. For present purposes, the term *hominin clade* encompasses our species and its ancestors back to our last common ancestor with the chimpanzee clade, around 6 million years ago. Bradley 2008 gives a useful overview.
4. For example, see White et al. 2009.
5. Bradley 2008.
6. Wallace 1898, 133.
7. Harcourt-Smith 2010.
8. For example, the position of foramen magnum and the morphology of the femur in *Sahelanthropus* and *Orrorin*: Richmond and Jungers 2008; Pickford 2005.
9. Grine et al. 2009.

10. Lesnik and Thackeray 2007.
11. Brain and Sillen 1988.
12. Ardrey 1963; Hart and Sussman 2009.
13. For example, see discussion in Pobiner et al. 2008.
14. Fleagle et al. 2010.
15. The ecology of mixed-species flocks is reviewed in detail by Harrison and Whitehouse 2011.
16. Camphuysen and Webb 1999.
17. For example, see Cords 1990.
18. King and Cowlishaw 2009.
19. Majolo and Ventura 2004.
20. Elder and Elder 1970.
21. Brain and Sillen 1988.
22. James 1989.
23. Caldararo 2002.
24. Two meticulous examples are Karkanas et al. 2007 on Qesem Cave, and Preece et al. 2006 on Beeches Pit.
25. Gowlett 2006.
26. For example, Mashkour et al. 2009.
27. For example, Lord et al. 2007.
28. Stewart 2004.
29. Stewart 2004, 181, table 1
30. Münzel and Conard 2004
31. Fiore et al. 2004.
32. Karavanić et al. 2008.
33. Karavanić et al. 2008, 269.
34. Bocheński and Tomek 2004; Tomek and Bocheński 2006.
35. Eastham 1997.
36. Fiore et al. 2004.
37. Bocheński et al. 2009.
38. Driver 1988; Fladmark et al. 1988.
39. Discussed well in Stewart 2004.
40. Lord et al. 2007.
41. Reviewed and summarized by Dorado et al. 2009.
42. A recent review of cultural developments in the Epipaleolithic is given by Richter et al. 2011.
43. Blockley and Pinhasi 2011.
44. Bar-Yosef 1988; Garrod and Bate 1957.
45. Rollefson 2008.
46. Maher et al. 2011; Tchernov and Valla 1997.
47. Wilson 2007.
48. Tchernov 1984; 1991a; 1993; and 1994.
49. Tchernov 1991b.
50. Wycoll and Tangri 1991.
51. Tchernov 1991b.
52. Weissbrod et al. 2005.
53. Nadel et al. 2008.
54. For more on spiny mice, see chapter 5.

55. Clutton-Brock 1995.
56. Tchernov and Valla 1997.
57. Gray et al. 2010.
58. Feddersen-Petersen 2008 discusses behaviors in dogs and wolves.
59. Cafazzo et al. 2010.
60. Driscoll and Macdonald 2010.
61. Moore et al. 2000, 14–17.
62. Bakke et al. 2009 gives a good overview and a specific case study of the evidence.
63. Usefully summarized in Watkins 2010, table 1. Also see Richter et al. 2011; Saña Segui et al. 1999.
64. Moore et al. 2000.
65. Hole 1984; Weisdorf 2005.
66. Yeshurun et al. 2009.
67. Twiss 2007, 140.
68. Arbuckle and Ozkaya 2006.
69. Starkovich and Stiner 2009.
70. Sapir-Hen et al. 2009.
71. Yeshurun et al. 2009.
72. Makarewicz 2009, 85.
73. Smith 1998 gives a good overview.
74. Piperno and Dillehay 2008.
75. Ranere et al. 2009.
76. Wang et al. 2010.
77. Denham et al. 2003.
78. Haberle 2007.
79. Dufraisse 2006.
80. Van der Mieroop 1999 for Mesopotamia; Chesson and Philip 2003 for the Levant.
81. Pigière and Henrotay (2011) attribute the distribution of camels in the northern Roman Empire to the excellent road system.
82. Of the many sources on Roman Britain, Mattingley 2007 is a useful introduction.
83. O'Connor 2010c.
84. Woodward et al. 1993.
85. Armitage 1994; Rielly 2010.
86. Maltby 2010.
87. O'Connor 1991, 257, table 70.
88. Rielly 2010, 139–40.
89. For example, Maltby 2010, 256–63, 272.
90. Maltby 2010, 274–77.
91. Parker 1988.
92. Maltby 2010, 303.
93. Serjeantson and Morris 2011.
94. Fox 1891.
95. Care over the burial of "pet" dogs extends into the past and across cultures. See Vellanoweth et al. 2008 for an excellent example.
96. King 2005.
97. For example, the Church Street Roman sewer in York; Buckland 1976.
98. Bintliff and Snodgrass 1988.

CHAPTER 4. MESOMAMMALS

1. The 29 October 1915 loss of the *Endurance* in pack ice made it necessary to shoot the ship's cat and three sledge-dog puppies; certain of the crew "seemed to feel the loss of their friends rather badly." Shackleton 1919, 81.
2. Felidae, 54–169, in Wilson and Mittermeier 2009.
3. Gentry et al. 2004.
4. For example, see Daniels et al. 2001 for interbreeding among feral cats and wildcats in Scotland.
5. O'Connor 2000b.
6. Bradshaw et al. 1999; Bradshaw 2006. Typical prey items are only 1 percent of the cat's body weight; Pearre and Maass 1998.
7. Harper 2005.
8. Pearre and Maass 1998.
9. Fitzgerald 1988.
10. Todd 1978.
11. Elton 1953; Fitzgerald 1988. My own cats show this behavior.
12. Driscoll et al. 2009, 9977–78.
13. Serpell 1988 and 1996; Watson and Weinstein 1993.
14. For example, in ancient Egypt; Malek 1993.
15. Liberg and Sandell 1988.
16. For example, Nicastro 2004. Character displacement occurs when the presence of a sympatric competitor species, usually closely related, causes a species to restrict or alter its behavior or even morphology to accentuate differences from the competitor species. See Losos 2000.
17. Cameron-Beaumont et al. 2002.
18. Driscoll et al. 2009.
19. Driscoll et al. 2009, 9975.
20. The consequences of this reticulate evolution is well summarized by Smouse 2000.
21. Tchernov 1991a and 1991b.
22. O'Connor 2007.
23. For example, see Meadow 1989.
24. Davis 1989.
25. Vigne et al. 2004.
26. "One of the potentially most reliable indicators of the presence of a domestic animal is finding its remains at a site in a region that is beyond the natural range of its wild relatives"; Meadow 1989, 84.
27. Zeuner 1963; Lowe 2005.
28. Zeuner 1963, 39.
29. Malek 1993, 45.
30. Malek 1993.
31. Zeuner 1963, 390; Serpell 1988, 152.
32. Kitchener and O'Connor 2010.
33. O'Connor 2007.
34. Adalsteinsson and Blumenberg 1983.
35. Long 2003, 311–15.
36. Nogales et al. 2004; Vázquez-Domínguez et al. 2004.
37. Bergstrom et al. 2009.
38. Slater and Shain 2005.

39. Dauphiné and Cooper 2009.
40. Van Heezik et al. 2010.
41. Baker et al. 2008.
42. Honeycutt 2010 reviews the copious and often contradictory literature on this matter.
43. Plug 1993.
44. Horsburgh 2008.
45. "Black-backed jackal," in Wilson and Mittermeier, eds. 2009, 419–20.
46. Gilbert-Norton et al. 2009.
47. Prange and Gehrt 2004.
48. Long 2003, 240–41.
49. Contesse et al. 2004.
50. Jankowiak et al. 2008.
51. Delibes-Matteos et al. 2008.
52. Red fox, in Wilson and Mittermeier 2009, 441–42.
53. Canidae, in Wilson and Mittermeier 2009, 352–411.
54. To my synesthesic nose, fox scent is khaki, while badger scent is a rich purple-brown, like fresh liver.
55. Finch 2007; Bevan 2010.
56. Kruuk 1972; Saunders et al. 2011.
57. Harris and Rayner 1986.
58. Battersby 2005.
59. *BBC News Magazine* online, Monday, 7 June 2010.
60. BBC Radio 4 *Today* programme, Monday, 7 June 2010.
61. Holly Williams, "We shouldn't cry wolf about foxes," *Independent (London)*, 15 June 2010, 13.
62. Evidence of the ability of cats to "see off" foxes is extensive (e.g., on YouTube), though mainly anecdotal. In an intriguing twist to the tale, cats may be a restricting factor in the conservation of urban populations of the San Joaquin kit fox; see Harrison et al. 2011.
63. For example, John Pugh, Welsh Farmers Fox Control Association, quoted at length by Andy McSmith, "In pursuit of London's Public Enemy no. 1," *Independent*, 12 June 2010, 23.
64. Williams 2010, "We shouldn't cry wolf."
65. König 2008.
66. König 2008, 104, table 2.
67. Robardet et al. 2011.
68. König 2008, 104, table 2.
69. Fairnell and Barrett 2007.
70. Personal communication with Eva Fairnell, to whom many thanks for discussing foxes and much else.
71. Evans 2010; Keene 1981.
72. Harris and Rayner 1986.
73. Tabor 1983.
74. Dewey and Fox 2001. "Procyon lotor," Animal Diversity Web, http://animaldiversity.ummz.umich.edu/site/accounts/information/Procyon_lotor.html (accessed 29 June 2010).
75. Prange et al. 2004.
76. Bozek et al. 2007.
77. Bartosiewicz et al. 2008.
78. Bartosiewicz et al. 2008, 294.

79. Prange et al. 2003.
80. Dawson et al. 2009.
81. Herr et al. 2009.
82. Otali and Gilchrist 2004.
83. Gehrt 2007.
84. Gehrt and Prange 2006.
85. Prange and Gehrt 2007.
86. Allen et al. 1989; Grayson 2001.
87. Grayson 2001, 18.
88. Merlo 2002. "Phalanger orientalis," Animal Diversity Web http://animaldiversity.ummz.umich.edu/site/accounts/information/Phalanger_orientalis.html (accessed 29 June 2010).
89. Hughes et al. 2008.
90. Mittermeier et al. 2007, "World's most endangered primates revealed," http://www.IUCN.org/?4753/worlds-most-endangered-primates-revealed (accessed 18 February 2010). This source adds that 48 percent of the world's 634 primate species are classified as "threatened."
91. See, as an introduction, Fuentes 2006: Agustin Fuentes is a key author on the relations between people and other primates.
92. Yamagiwa and Hill 1983.
93. Chauhan and Pirta 2010.
94. Waite et al. 2007.
95. Waite et al. 2007, 284.
96. Devi and Saikia 2008.
97. Lee and Priston 2005.
98. Lee and Priston 2005, 2; Knight 1999.
99. YouTube is a rich source of examples, notably from Knowsley Safari Park, England, where the baboons seem to understand cars better than the visitors understand baboons.
100. Warren 2008, 2.
101. Warren 2008, 4–5, table 1.
102. Naughton-Treves et al. 1998; McKinney 2010.
103. Vijman and Nekaris 2010.

CHAPTER 5. RATS, MICE, AND OTHER RODENTS

1. Long 2003 devotes a lengthy catalog to 67 species of introduced rodents, of which perhaps 10 have been redistributed widely or established substantial populations.
2. For non-UK readers, Roland Rat was a cartoon and puppet character that became the public face of the newly launched TV Channel 4. The far-sighted network controller who backed the introduction of Roland Rat is currently (2012) the chancellor of the university that employs the author.
3. Marshall and Sage 1981; Suzuki et al. 2004; Macholán 2006.
4. Schwarz and Schwarz 1943.
5. Auffray et al. 1988; Cucchi et al. 2005; Cucchi and Vigne 2006; Boursot et al. 1993 review the prehistory of the group in detail, according to the state of knowledge at that time. Subsequent research has refined the detail but not overturned the main conclusions.

6. Willcox et al. 2009.
7. Boursot et al. 1993, 136–37; Bonhomme et al. 2011.
8. Cucchi 2008. Elliptical Fourier Transform shape analysis on the first molar confirms this specimen to be *M. m. domesticus.*
9. Morales Muñiz et al. 1995.
10. Boursot et al. 1993.
11. Coy 1984; Harcourt 1979. The history of house mice in Britain is dealt with in O'Connor 2010c.
12. O'Connor 1986.
13. Searle et al. 2009. Searle's pursuit of mouse genetics across the former British Empire has been an inspiration to behold; see also Bonhomme et al. 2011.
14. Gyllensten and Wilson 1987; Prager et al. 1992.
15. Long 2003, 208–9.
16. Tichy et al. 1994.
17. For example, the population studied in depth by Waudby (2010).
18. Bergstrom et al. 2009.
19. Berry 1970; Corbet and Harris 1991.
20. "All mice must die sometime"; Berry 1970, 235.
21. Ganem and Searle 1996; Frynta et al. 2005.
22. Natoli and DeVito 1991.
23. Dagg 2011.
24. Fitzwater 1970.
25. Drummond 1992.
26. Fitzwater 1970.
27. Hinton 1930, 48.
28. Dawson 1924.
29. Podzorski 1990, 88–89.
30. In Culpeper's *Pharmacopeia Londonensis*, 6th ed., 1659.
31. Discussed but not sourced in Berry 1970, 221–22.
32. Tucker et al. 1992.
33. Leach and Main 2008. Part of their modus operandi was a questionnaire survey—presumably not addressed to the mice?
34. Balcombe 2010, 78.
35. Blanchard 2010: "The pattern of disconnections between data and conclusions is so pervasive as to demolish the scientific value of the exposition." Ouch.
36. Long 2003, 201–2.
37. Novákova et al. 2008.
38. Long 2003, 188–90.
39. Ervynck 2002, 98.
40. Long 2003, 198–99.
41. Toškan and Kryštufek 2006.
42. Rielly 2010.
43. O'Connor 1992.
44. Armitage 1994.
45. Rielly 2010.
46. Rielly 2010, 143.

47. Numerous sources contest this point: see Davis 1986; Antoine 2008; Welford and Bossack 2009 and 2010.
48. My thanks to Bruce Campbell, Queens University Belfast, for his views on this point.
49. Haensch et al. 2010.
50. Armitage 1993.
51. Smith 2001.
52. Bratten 1995.
53. Pavao-Zuckerman 2007.
54. Landon 2009.
55. Milne and Crabtree 2001.
56. Reitz and Honerkamp 1983.
57. Van Schalkwyk et al. 1995.
58. Voigt and von den Driesch 1984.
59. Voigt and von den Driesch 1984, 100.
60. I am grateful to Thomas Biginagwa, University of Dar Es-Salaam, for permission to quote results from his recent excavations on the Pangani.
61. Juma 2004; Horton 1996. I am grateful to Stephanie Wynn-Jones for references to East African material.
62. Dieterlen 1975, 357.
63. Silver 1927.
64. Dieterlen, 1975, 357.
65. Myers and Armitage 2004, "Rattus norvegicus," Animal Diversity Web, http://animaldiversity.ummz.umich.edu/site/accounts/information/Rattus_norvegicus.html (accessed 30 June 2010).
66. Silver 1927.
67. Dieterlen 1975, 358–60.
68. McGuire et al. 2006.
69. Traweger et al. 2006.
70. Traweger et al. 2006, 124.
71. Traweger et al. 2006, 114.
72. Langton et al. 2001.
73. Langton et al. 2001, 706; Childs 1986.
74. Battersby et al. 2002; Fink 2006.
75. BBC News Online, January 2007, "Rat number surge is health risk," http://news.bbc.co.uk/1/hi/6234893.stm (accessed 30 June 2010); *Irish Times* online, 2010, "Wild in the city," http://www.irishtimes.com/timeseye/wild/article_p3b.htm (accessed 30 June 2010).
76. Long 2003, 80–83.
77. Matisoo-Smith and Robins 2009.
78. Barnes et al. 2006.
79. Matisoo-Smith and Robins 2004.
80. Braithwaite 1980.
81. Veitch 2002.
82. Veitch 2002, 355.
83. Okubo et al. 1989; Lever 2009, 31–33.
84. Long 2003.
85. Sheail 1999.
86. Long 2003, 146.

87. Usher et al. 1992.
88. Cooper et al. 2008.

CHAPTER 6. BIRDS

1. For "othering," see Fabian 1990; Bowman 2008.
2. Watson 1975; Long 1981, 211.
3. Long 1981, 210.
4. Long 1981, 208–9; Carl and Guiguet 1972.
5. Pliny, *Natural History* 17:6; Varro, *On Agriculture*, 1:38; Columella, *De Re Rustica*, 2:14; Atwell 1996
6. Husselman 1953.
7. Chambers 1920.
8. Ramsay and Tepper 2010.
9. Amirkhani et al. 2010.
10. İmamoğlu et al. 2005.
11. Dixon 1851, 2.
12. Murton et al. 1972a; 1972b.
13. Murton et al. 1972a, 848, table 6.
14. Murton et al. 1972b, 883.
15. Sol et al. 2000.
16. "Save the Trafalgar Square pigeons," http://www.savethepigeons.org/background.html (accessed 5 July 2010); "Soaring cost of hawks to scare pigeons from Trafalgar Square," *London Today*, http://www.thisislondon.co.uk/standard/article-23746422 (accessed 5 July 2010).
17. Murton et al. 1972a, 842.
18. Rose et al. 2006.
19. Miyata and Fujita 2008.
20. Buijs and van Wijnen 2001.
21. Shaw et al. 2008.
22. See, for example, numerous online ads for pigeon control by hawks and other means: e.g., "Pigeon Control London," http://www.businessmagnet.co.uk/company/pigeoncontrollondon-135862.htm (consulted 15 February 2012). The hawks occasionally take less acceptable prey such as small dogs; http://www.telegraph.co.uk/earth/earthnews/8505777/Hawks-hunt-down-handbag-dogs-in-central-London.html (consulted 15 February 2012).
23. "Industry Statistics Sampler NAICS 561710," http://www.census.gov.uk/epcd/ec97/industry/E561710.htm (accessed 5 July 2010).
24. Haag-Wackernagel and Geigenfeind 2008.
25. Haag-Wackernagel 2006.
26. Haag-Wackernagel and Geigenfeind 2008, 715–16.
27. Majewska et al. 2009.
28. Haag-Wackernagel and Geigenfeind 2008, 719–20.
29. Jerolmack 2008.
30. Sadly, a full discussion of the works of Tom Lehrer is beyond the remit of this book, but comparison of "Poisoning Pigeons" with other works such as "Be Prepared," which spoofs the Boy Scouts,

and "Vatican Rag," which is self-explanatory, indicates a preparedness to assault fondly regarded institutions, including feeding pigeons (and squirrels in a later verse).

31. Summa et al. 2012, Tarsitano et al. 2010, Daniels et al. 2009 give a thorough review of the prevalence, outcomes, and costs of dog bites to children in the United States.
32. Morales Muñiz et al. 1995.
33. Estabrook 1907 gives a contemporary perspective on sparrows as pests, noting with satisfaction that some people "are doing honest work toward sparrow extermination"; Brodhead 1971 outlines the career of one such exterminator.
34. Keeping sparrows as caged pets still has its adherents: for example, Jennifer Patterson, "How to raise house sparrows," at http://www.ehow.com/how_8773780_raise-house-sparrows.html (consulted 15 February 2012). She observes that "The house sparrow makes an affable pet."
35. Long 1981, 378; Le Souef 1958.
36. Martin and Fitzgerald 2005; Sol et al. 2002.
37. Ericson et al. 1997.
38. Chamberlain et al. 2007; Shaw et al. 2008.
39. Potts 1967.
40. Clergeau and Quenot 2007.
41. Lever 2009, 186. Parakeet myths have otherwise escaped the academic literature, but are regularly revisited by the popular press. See, for example, *Mail Online*, www.dailymail.co.uk/news/article-1217403/Open-season-parakeets-exotic-birds-shot-licence-new-regulations.html.
42. Lever 2009, 180; Butler 2003; 2005. Even odder than the parakeet myth is the fact that Hendrix played a gig in the small Yorkshire town of Ilkley in 1967. Something of the character of Ilkley is shown by the fact that older locals still recall the event, including the police constable who was sent to close it down following complaints about the noise.
43. Butler 2002.
44. Strubbe et al. 2010.
45. Jones and Reynolds 2008 mention competition at feeders in passing. Butler 2005 makes the point that some parakeet populations appear to make little use of food put out by people, and that the spread into rural areas away from handouts indicates that the birds are not sustained by supplementary feeding by people.
46. Rock 2005, 338.
47. Rock 2005; Parslow 1967.
48. Rock 2005, 341; Skórka et al. 2006.
49. Soldatini et al. 2008.
50. Maciusik et al. 2010; Indykiewicz 2005.
51. Seed et al. 2009.
52. Emery 2005; Emery and Clayton 2004; Bentley-Condit and Smith 2010.
53. Bugnyar and Heinrich 2006.
54. Joshua Klein, "A vending machine for crows," http://www.josh.is/projects/crows/crows_JoshKlein.pdf (accessed 5 July 2010).
55. White 2005.
56. Gade 2010.
57. Gadgill 2001.
58. Marzluff and Angell 2005.
59. Peters and Schmidt 2004.
60. Driver 1999.

61. Bond and O'Connor 1999, 392–93.
62. O'Connor 1991, 260–62.
63. Evans et al. 1999. Sadly, the reestablishment of red kites in England is not without its opponents. Poisoned bait continues to be illegally placed in likely feeding areas: for a typical case, see http://www.ilkleygazette.co.uk/news/news_local/8987114.Red_kite_chicks_found_poisoned_near_Ilkley/.
64. Vuorisalo et al. 2003.
65. Vuorisalo et al. 2003, 75–76.
66. Vuorisalo et al. 2003, 80–81.

CHAPTER 7. COMMENSALISM, COEVOLUTION, AND CULTURE

1. McKinney 2006; Heinsohn 2001 and 2003.
2. Harris and Rayner 1986.
3. See chap. 6.
4. Flinn 1997; Castro and Toro 2004.
5. Yeshurun et al. 2009.
6. Harris 1977; Baker et al. 2007.
7. Lukasz 2001; DeCandido and Allen 2006.
8. Reviewed in Haslam et al. 2009; also McGrew 2004; van Schaik et al. 2003 for culture in orangutans.
9. McGrew 2010.
10. Mannu and Ottoni 2009; Ottoni and Izar 2008.
11. Nihei and Higuchi 2001; Emery and Clayton 2004.
12. Zhang et al. 2007.
13. For example, langurs during times of drought in India; Waite et al. 2007.
14. Chap. 6; also Shaw et al. 2008.
15. By association of words and action, the toddler son of friends learned that the family cat was called "Gently . . . ."
16. Peresani et al. 2011.
17. Bradshaw and Paul 2010.
18. Mithen 1999.
19. Bradshaw and Paul 2010.
20. Wilson 1984.
21. Kellert and Wilson 1993, 454.
22. Kellert and Wilson 1993.
23. Kellert and Wilson 1993, 203.
24. Kellert and Wilson 1993, 203.
25. Joye and De Block 2011.
26. Levy 2003.
27. Beck and Katcher 2003.
28. Herzog 2002.
29. My thanks to Lisa-Marie Shillitoe for this information.
30. Webster 2007.
31. Crooks and Soulé 1999.
32. Piper et al. 2012.

CHAPTER 8. PLANNING FOR THE FUTURE

1. Dick 1968.
2. Kahn et al. 2009.
3. Antonioli and Reveley 2005.
4. Cooper 1976; Antonioli and Reveley 2005.
5. Rice et al. 1973.
6. Gullone 2000.
7. Pyle 1993; Louv 2008; Maller et al. 2010; Cookson 2011.
8. Hounslow et al. 2010.
9. Beatley 2009; Miller 2008; Delavari-Edalat and Abdi 2010.
10. Beatley 2009, 215.
11. Beatley 2009, 235.
12. Webster 2007.
13. Rutz 2008.
14. Berry 1987; Paulos et al. 2008; Cooper et al. 2007.
15. Jerolmack 2008.

Bibliography

Adalsteinsson, S., and B. Blumenberg. 1983. Possible Norse origin for two northeastern United States cat populations. *Zeitschrift für Tierzuchtung und Zuchtungsbiologie* 100: 161–74.

Alberti, M., J. M. Marzluff, E. Shulenberger, G. Bradley, C. Ryan, and C. ZumBrunnen. 2008. Integrating humans into ecology: Opportunities and challenges for studying urban ecosystems. In *Urban Ecology: An International Perspective on the Interaction between Humans and Nature*, ed. J. Marzluff, E. Schulenberger, W. Endlicher, M. Alberti, G. Bradley, C. Ryan, C. ZumBrunnen, and U. Simon, 143–58. New York: Springer Science and Business Media.

Allen, J., C. Gosden, and J. P. White. 1989. Human Pleistocene adaptations in the Tropical Island Pacific: Recent evidence from New Ireland, a greater Australian outlier. *Antiquity* 63: 548–61.

Allison, P. A., and R. Bottjer, eds. 2011. *Taphonomy: Process and Bias through Time.* New York: Springer.

Amirkhani, A., H. Okhovat, and E. Zamani. 2010. Ancient pigeon houses: Remarkable example of the Asian culture crystallised in the architecture of Iran and Central Anatolia. *Asian Culture and History* 2 (2): 45–57.

Anderson, A. 1989. *Prodigious Birds: Moas and Moa Hunting in Prehistoric New Zealand.* Cambridge: Cambridge University Press.

Andrezjewski, R., J. Babińska-Werka, J. Gliwicz, and J. Goszczyński. 1978. Synurbization processes in an urban population of *Apodemus agrarius.* Part 1, Characteristics of population in urbanization gradient. *Acta theriologica* 23: 341–58.

Antoine, D. 2008. The archaeology of "plague." *Medical History Supplement* 27: 101–14.

Antonioli, C., and M. A. Reveley. 2005. Randomised controlled trial of animal facilitated therapy with dolphins in the treatment of depression. *British Medical Journal* 331: 1231–34.

Arbuckle, D., and V. Ozkaya. 2006. Animal exploitation at Körtik Tepe: An early Aceramic Neolithic site in southeastern Turkey. *Paléorient* 32 (2): 113–36.

Ardrey, R. 1963. *African Genesis: A Personal Investigation into the Animal Origins and Nature of Man.* New York: Delta Books.

Armitage, P. L. 1993. Commensal rats in the New World, 1492–1992. *Biologist* 40: 174–78.

Armitage, P. L. 1994. Unwelcome companions: Ancient rats reviewed. *Antiquity* 68: 231–40.

Arnold, P. G., J. P. Sadler, M. O. Hill, A. Pullin, S. Rushton, K. Austin, E. Small, B. Ward, R. Wadsworth, R. Sanderson, and K. Thompson. 2006. Biodiversity in urban habitat patches. *Science of the Total Environment* 360: 196–204.

Atwell, D. 1996. A bird in the dovecote is worth two in the bush. *Avicultural Magazine* 102 (4): 169–73.

Auffray, J.-C., E. Tchernov, and E. Nevo. 1988. Origine du commensalisme de la souris domestique *Mus musculus domesticus* vis-à-vis l'homme. *Comptes Rendus de l'Académie des Sciences de Paris* 307: 517–22.

Baker, P. J., C. V. Dowding, S. E. Molony, P. C. L. White, and S. Harris. 2007. Activity patterns of urban red foxes (*Vulpes vulpes*) reduce the risk of traffic-induced mortality. *Behavioural Ecology* 18: 716–24.

Baker, P. J., S. E. Molony, E. Stone, I. C. Cuthill, and S. Harris. 2008. Cats about town: Is predation by free-ranging cats *Felis catus* likely to affect urban bird populations? *Ibis* 150: 86–99.

Bakke, J., Ø. Lie, E. Heegard, T. Dokken, G. H. Haug, H. H. Birks, P. Dulski, and T. Nilsen. 2009. Rapid oceanic and atmospheric changes during the Younger Dryas cold period. *Nature Geoscience* 2: 202–5.

Balcombe, J. 2010. Laboratory rodent welfare: Thinking outside the cage. *Journal of Applied Animal Welfare Science* 13 (1): 77–88.

Banks, A. N., Q. P. Humphrey, Rachel Coombes Crick, Stuart Benn, Derek A. Ratcliffe, and Elizabeth M. Humphreys. 2010. The breeding status of Peregrine Falcons Falco peregrinus in the UK and Isle of Man in 2002. *Bird Study* 57 (4): 421–36.

Barnes, S. S., E. Matisoo-Smith, and T. L. Hunt. 2006. Ancient DNA of the Pacific rat (Rattus exulans) from Rapa Nui (Easter Island). *Journal of Archaeological Science* 33: 1536–40.

Bar-Oz, G., and Y. Tepper. 2010. Out on the tiles: Animal footprints from the Roman site of Kefar 'Othnay (Legio), Israel. *Near Eastern Archaeology* 73 (4): 244–47.

Bartosiewicz, M., H. Okarma, A. Zalewski, and J. Szczęsna. 2008. Ecology of the raccoon (*Procyon lotor*) from western Poland. *Annales Zoologica Fennici* 45: 291–98.

Bar-Yosef, O. 1988. The Natufian culture in the Levant: Threshold to the origins of agriculture. *Evolutionary Anthropology* 6 (5): 159–77.

Battersby, J. 2005. *UK Mammals: Species Status and Population Trends.* Joint Nature Conservation Committee, Peterborough, UK. http://www.jncc.gov.uk/page-1829.

Battersby, S., R. Parsons, and J. P. Webster. 2002. Urban rat infestations and the risk to public health. *Journal of Environmental Health Risks* 1 (2): 1–11.

Beatley, T. 2009. Biophilic urbanism: Inviting nature back into our communities and into our lives. *William and Mary Environmental Law and Policy Review* 34 (1): 209–38.

Beck, A. M., and A. H. Katcher. 2003. Future directions in human-animal bond research. *American Behavioral Scientist* 47: 79–93.

Bell, T. 1837. *A History of British Quadrupeds.* London: John van Voorst.

Benítez-López, A., R. Alkemade, and P. A. Vermeij. 2010. The impact of roads and other infrastructure on mammal and bird populations: A meta-analysis. *Biological Conservation* 143: 1307–16.

Bentley-Condit, V. K., and E. O. Smith. 2010. Animal tool-use: Current definitions and an updated comprehensive catalog. *Behaviour* 147 (2): 185–221.

Bergstrom, D. M., A. Lucieer, K. Kiefer, J. Wasley, L. Belbin, T. K. Pedersen, and S. L. Chown. 2009. Indirect effects of invasive species removal devastates World Heritage Island. *Journal of Applied Ecology* 46 (1): 73–81.

Berry, R. J. 1970. The natural history of the house mouse. *Field Studies* 3: 219–62.

Berry, R. J. 1987. Scientific natural history: A key base to ecology. *Biological Journal of the Linnaean Society* 32: 17–29.

Bevan, J. 2010. Agricultural change and the development of hunting in the eighteenth century. *Agricultural History Review* 58 (1): 49–75.

Bintliff, J., and A. M. Snodgrass. 1988. Off-site pottery distributions: A regional and interregional perspective. *Current Anthropology* 29 (3): 506–13.

Blanchard, R. J. 2010. Animal welfare beyond the cage . . . and beyond the evidence? *Journal of Applied Animal Welfare Science* 13 (1): 89–95.

Bleed, P. 2006. Living in the human niche. *Evolutionary Anthropology* 15: 8–10.

Blockley, S., and R. Pinhasi. 2011. A revised chronology for the adoption of agriculture in the Southern Levant and the role of Lateglacial climatic change. *Quaternary Science Reviews* 30 (1–2): 98–108.

Bocheński, Z., and T. Tomek. 2004. Bird remains from a rock-shelter in Kruczka Skala (Central Poland). *Acta zoologica cracowiensia* 47 (1–2): 27–47.

Bocheński, Z., T. Tomek, J. Wilczyński, J. Svoboda, K. Wertz, and P. Wojtal. 2009. Fowling during the Gravettian: The avifauna of Pavlov I, the Czech Republic. *Journal of Archaeological Science* 36: 2635–65.

Boivin, N., R. Blench, and D. Q. Fuller. 2009. Archaeological, linguistic and historical sources on ancient seafaring: A multidisciplinary approach to the study of early maritime contact and exchange in the Arabian Peninsula. In *The Evolution of Human Populations in Arabia*, ed. M. D. Petraglia and J. I. Rose, 251–78. New York: Springer Science and Business.

Bonanni, R., S. Cafazzo, C. Fantini, D. Pontier, and E. Natoli. 2007. Feeding-order in an urban feral domestic cat colony: Relationship to dominance rank, sex and age. *Animal Behaviour* 74 (5): 1369–79.

Bond, J. M., and T. P. O'Connor. 1999. *Bones from Medieval Deposits at 16–22 Coppergate and Other Sites in York*. Archaeology of York 15/5. York: Council for British Archaeology.

Bonhomme, F., A. Orth, T. Cucchi, H. Rajabi-Mahan, J. Catalan, P. Boursot, J.-C. Auffray, and J. Britton-Davidian. 2011. Genetic differentiation of the house mouse around the Mediterranean basin: Matrilineal footprints of early and late colonisation. *Proceedings of the Royal Society of London B* 278 (1708): 1034–43.

Boucher, D. H., S. James, and K. H. Keeler. 1982. The ecology of mutualism. *Annual Review of Ecology and Systematics* 13: 315–47.

Boursot, P., J.-C. Auffray, J. Britton-Davidian, and F. Bonhomme. 1993. The evolution of house mice. *Annual Review of Ecology and Systematics* 24: 119–52.

Bowman, G. 2008. Beyond othering. *American Ethnologist* 30 (4): 500–501.

Bozek, C. K., S. Prange, and S. D. Gehrt. 2007. The influence of anthropogenic resources in multi-scale habitat selection by raccoons. *Urban Ecosystems* 10: 413–25.

Bradley, B. J. 2008. Reconstructing phylogenies and phenotypes: A molecular view of human evolution. *Journal of Anatomy* 212 (4): 337–53.

Bradshaw, J. W. S. 2006. The evolutionary basis for the feeding behaviour of domestic dogs (*Canis familiaris*) and cats (*Felis catus*). *Journal of Nutrition* 136: 1927–31.

Bradshaw, J. W. S., G. F. Horsfield, J. A. Allen, and I. H. Robinson. 1999. Feral cats: Their role in the population dynamics of *Felis catus*. *Applied Animal Behaviour Science* 65: 273–83.

Bradshaw, J.W.S., and E. S. Paul. 2010. Could empathy for animals have been an adaptation in the evolution of *Homo sapiens*? *Animal Welfare* 19 (5): 107–12.

Brain, C. K., and A. Sillen. 1988. Evidence from the Swartkrans Cave for the earliest use of fire. *Nature* 336: 464–66.

Braithwaite, R. W. 1980. The ecology of *Rattus lutreolus* III: The rise and fall of a commensal population. *Australian Wildlife Research* 7: 199–215.

Bratten, J. R. 1995. Olive pits, rat bones, and leather shoe soles: A preliminary report on the organic remains from the Emanuel Point Shipwreck, Pensacola, Florida. In *Underwater Archaeology: Proceedings from the Society for Historical Archaeology Conference* 1995, ed. P. Johnson, 49–53. Washington, DC: Society for Historical Archaeology.

Brodhead, M. J. 1971. Elliot Coues and the Sparrow War. *New England Quarterly* 44 (3): 420–32.

Brown, T., and K. Brown. 2011. *Biomolecular Archaeology: An Introduction*. Chichester, UK: Wiley-Blackwell.

Buczacki, S. 2002. *Fauna Britannica*. London: Hamlyn.

Buckland, P. C. 1976. *The Environmental Evidence from the Church Street Roman Sewer System*. The Archaeology of York 14/1. London: Council for British Archaeology.

Buckley, M., S. Whitcher Kansa, S. Howard, S. Campbell, J. Thomas-Oates, and M. Collins. 2010. Distinguishing between archaeological sheep and goats using a single collagen peptide. *Journal of Archaeological Science* 37 (1): 13–20.

Budiansky, S. 1992. *The Covenant of the Wild: Why Animals Chose Domestication*. London: Weidenfeld & Nicholson.

Bugnyar, T., and B. Heinrich. 2006. Pilfering ravens, *Corvus corax*, adjust their behaviour to social context and identity of competitors. *Animal Cognition* 9: 369–76.

Buijs, J., and J. H. van Wijnen. 2001. Survey of feral rock doves (*Columba livia*) in Amsterdam: A bird-human association. *Urban Ecosystems* 5 (4): 235–41.

Butler, C. J. 2002. Breeding parrots in Britain. *British Birds* 95: 34–38.

Butler, C. J. 2003. Population biology of the introduced rose-ringed parakeet *Psittacula krameri* in the UK. Unpublished PhD thesis, University of Oxford.

Butler, C. J. 2005. Feral parrots in the continental United States and United Kingdom: Past, present, and future. *Journal of Avian Medicine and Surgery* 19 (2): 142–49.

Cafazzo, S., P. Valsecchi, R. Bonanni, and F. Natoli. 2010. Dominance in relation to age, sex and competitive contexts in a group of free-ranging domestic dogs. *Behavioural Ecology* 21 (3): 443–55. doi: 10.1093/beheco/arq001.

Caldararo, N. 2002. Human ecological intervention and the role of forest fires in human ecology. *Science of the Total Environment* 292: 141–65.

Cameron-Beaumont, C., S. E. Lowe, and J. W. S. Bradshaw. 2002. Evidence suggesting preadaptation to domestication throughout the small Felidae. *Biological Journal of the Linnaean Society* 75: 361–66.

Camphuysen, C. J., and A. Webb. 1999. Multi-species feeding associations in North Sea seabirds: Jointly exploiting a patchy environment. *Ardea* 87 (2): 177–98.

Carl, G. C., and C. J. Guiguet. 1972. *Alien animals in British Columbia*. Handbook 14, 2nd ed. Victoria, BC: British Columbia Provincial Museum.

Cassidy, R., and M. Mullin. 2007. *Where the Wild Things Are Now*. New York: Berg.

Castro, L., and M. A. Toro. 2004. The evolution of culture: From primate social learning to human culture. *Proceedings of the National Academy of Sciences* 101 (27): 10235–40.

Chamberlain, D. E., M. P. Toms, R. Cleary-McHarg, and A. M. Banks. 2007. House sparrow (*Passer domesticus*) habitat use in urbanised landscapes. *Journal of Ornithology* 148: 453–62.

Chambers, C. D. 1920. Romano-British dovecots. *Journal of Roman Studies* 10: 189–93.

Chauhan, A., and R. S. Pirta. 2010. Socio-ecology of two species of non-human primates, Rhesus Monkey (*Macaca mulatta*) and Hanuman Langur (*Semnopithecus entellus*) in Shimla, Himachal Pradesh. *Journal of Human Ecology* 30 (3): 171–77.

Chesson, M., and G. Philip. 2003. Tales of the city. *Journal of Mediterranean Archaeology* 16 (1): 3–16.

Childs, J. E. 1986. Size-dependent predation on rats (*Rattus norvegicus*) by house cats. *Journal of Mammalogy* 67: 196–99.

Clergeau, P., and F. Quenot. 2007. Roost selection flexibility of European starlings aids invasion of urban landscapes. *Landscape and Urban Planning* 80: 56–62.

Clutton-Brock, J. 1995. Origins of the dog: Domestication and early history. In *The Domestic Dog: Its Evolution, Behaviour, and Interactions with People*, ed. J. Serpell, 7–20. Cambridge: Cambridge University Press.

Clutton-Brock, J. 1999. *A Natural History of Domesticated Mammals.* 2nd ed. Cambridge: Cambridge University Press.

Clutton-Brock, J. 2012. *Animals as Domesticates: A World View through History*. East Lansing: Michigan State University Press.

Conard, N. J. 2003. Palaeolithic ivory sculptures from southwestern Germany and the origins of figurative art. *Nature* 426 (18/25): 830–32.

Contesse, P., D. Haggelin, S. Gloor, F. Bontadina, and P. Deplazes. 2004. The diet of urban foxes, (*Vulpes vulpes*) and the availability of anthropogenic food in the city of Zurich, Switzerland. *Mammalian Biology* 69: 81–95.

Cookson, L. J. 2011. A definition for wildness. *Ecopsychology* 3 (3): 187–93.

Cooper, C. A., A. J. Nett, D. P. Poon, and G. R. Smith. 2008. Behavioural responses of eastern gray squirrels in suburban habitats differing in human activity levels. *Northeastern Naturalist* 15 (4): 619–25.

Cooper, C. B., J. Dickinson, T. Phillips, and R. Bonney. 2007. Citizen science as a tool for conservation in residential ecosystems. *Ecology and Society* 12 (2): 1–11.

Cooper, J. E. 1976. Pets in hospitals. *British Medical Journal* 1: 698–700.

Corbet, G. B., and S. Harris. 1991. *The Handbook of British Mammals*. Oxford: Blackwell Scientific.

Cords, M. 1990. Vigilance and mixed-species associations of some East African forest monkeys. *Behavioural Ecology and Sociobiology* 26: 297–300.

Coy, J. P. 1984. The small mammals and amphibians. In *Danebury: An Iron Age Hillfort in Hampshire*, vol. 2, *The excavations 1968–1978: The Finds*, ed. B. Cunliffe, 526–31. London: CBA.

Crooks, K. R., and M. Soulé. 1999. Mesopredator release and avifaunal extinctions in a fragmented ecosystem. *Nature* 400: 563–66.

Crutzen, P. J. 2002. Geology of mankind. *Nature* 415: 23.

Cucchi, T. 2008. Uluburun shipwreck stowaway house mouse: Molar shape analysis and indirect clues about the vessel's last journey. *Journal of Archaeological Science* 35: 2953–59.

Cucchi, T., J.-D. Vigne, and J.-C. Auffray. 2005. First occurrence of the house mouse (*Mus musculus domesticus* Schwarz and Schwarz, 1943) in the Western Mediterranean: A zooarchaeological revision of subfossil occurrences. *Biological Journal of the Linnaean Society* 84: 429–45.

Cucchi, T., and J. D. Vigne. 2006. Origin and diffusion of the house mouse in the Mediterranean. *Journal of Human Evolution* 21 (2): 95–106.

Curthoys, A., and J. Docker. 2010. *Is History Fiction?* 2nd ed. Sydney: University of New South Wales Press.

Dagg, J. L. 2011. Exploring mouse trap history. *Evolution: Education and Outreach* 4: 397–414. doi: 10.1007/s12052-011-0315-8.

Daniels, D. M., R. Ritzi, and J. O'Neil. 2009. Analysis of nonfatal dog bites in children. *Journal of Trauma-Injury Infection and Critical Care* 66 (3): S17–S22. doi: 10.1097/TA.0b013e3181937925.

Daniels, M. J., M. A. Beaumont, P. J. Johnson, D. Balharry, D. W. Macdonald, and E. Barratt. 2001. Ecology and genetics of wild-living cats in the north-east of Scotland and the implications for the conservation of the wildcat. *Journal of Applied Ecology* 38: 146–61.

Dauphiné, N., and R. J. Cooper. 2009. Impacts of free-ranging domestic cats (*Felis catus*) on birds in the United States: A review of recent research with conservation and management recommendations. In *Proceedings of the Fourth International Partners in Flight Conference: Tundra to Tropics*, 205–19. http://www.partnersinflight.org/pubs/McAllenProc/articles/PIF09_Anthropogenic%20 Impacts/Dauphine_1_PIF09.pdf.

Davis, D. E. 1986. The scarcity of rats and the Black Death: An ecological history. *Journal of Interdisciplinary History* 16 (3): 455–70.

Davis, S. J. M. 1989. Some more animal remains from the Aceramic Neolithic of Cyprus. In *Fouilles Récentes à Khirokitia (Chypre) 1983–1986*, ed. D. A. L. Brun, 189–221. Paris: Editions Recherche sur les Civilisations.

Dawson, J., M. Huck, R. J. Delahay, and T. J. Roper. 2009. Restricted ranging behaviour in a high-density population of urban badgers. *Journal of Zoology* 277 (1): 45–53.

Dawson, W. R. 1924. The mouse in Egyptian and later medicine. *Journal of Egyptian Archaeology* 10 (2): 83–86.

DeCandido, R., and D. Allen. 2006. Nocturnal hunting by peregrine falcons at the Empire State Building, New York City. *Wilson Journal of Ornithology* 118 (1): 53–58.

Delavari-Edalat, F., and M. R. Abdi. 2010. Human-environment interactions based on biophilia values in an urban context: Case study. *Journal of Urban Planning and Development* 136 (2): 162–68.

Delibes-Matteos, M., J. Fernandez de Simon, R. Villafuerte, and P. Ferreras. 2008. Feeding responses of the red fox (*Vulpes vulpes*) to different wild rabbit (*Oryctolagus cuniculus*) densities: A regional approach. *European Journal of Wildlife Research* 54: 71–78.

De Mazancourt, C., M. Loreau, and U. Dieckmann. 2004. Understanding mutualism when there is adaptation to the partner. *Journal of Ecology* 93: 305–14. doi: 10.1111/j.1365-2745.2004.00952.x.

Denham, T. P., S. G. Haberle, C. Lentfer, R. Fullagar, J. Field, M. Therin, N. Proch, and B. Winsborough. 2003. Origins of agriculture at Kuk Swamp in the Highlands of New Guinea. *Science* 301 (5630): 189–93.

Devi, O. S., and P. K. Saikia. 2008. Human-monkey conflict: A case study at Gauhati University Campus, Jalukbari, Kamrup, Assam. *Zoos' Print* 23 (2): 15–18.

Dick, P. K. 1968. *Do Androids Dream of Electric Sheep?* New York: Ballantine Books.

Dickman, C. 1992. Commensal and mutualistic interactions among terrestrial vertebrates. *Trends in Ecology and Evolution* 7: 194–97.

Dieterlen, F. 1975. Bamboo rats, mole rats and murid rodents. In *Grzimek's Animal Life Encyclopaedia*, ed. B. Grzimek, 346–87. New York: Van Nostrand Rheinhold.

Dixon, E. S. 1851. *The Dovecote and the Aviary: Being Sketches of the Natural History of Pigeons and Other Birds in a Captive State, With Hints for Their Management.* London: John Murray.

Dorado, G., I. Rey, F. J. Sánchez-Cañete, F. Luque, I. Jiménez, M. Gálvez, A. Sáiz, and V. F. Vásquez. 2009. Ancient DNA to decipher the domestication of the dog. *Arqueobios* 3 (1): 127–32.

Douglas, I., and J. P. Sadler. 2011. Urban wildlife corridors: Conduits for movement or linear habitat? In *The Routledge Handbook of Urban Ecology*, ed. R. Wong, 274–86. London: Routledge.

Driscoll, C. A., and D. W. Macdonald. 2010. Top dogs: Wolf domestication and wealth. *Journal of Biology* 9: 10.

Driscoll, C. A., D. W. Macdonald, and S. J. O'Brien. 2009. From wild animals to domestic pets, an evolutionary view of domestication. *Proceedings of the National Academy of Science* 106 (supp. 1): 9971–78.

Driver, J. C. 1988. Late Pleistocene and Holocene vertebrates and palaeoenvironments from Charlie Lake Cave, northeast British Columbia. *Canadian Journal of Earth Sciences* 25 (10): 1545–53.

Driver, J. C. 1999. Raven skeletons from Palaeoindian contexts, Charlie Lake Cave, British Columbia. *American Antiquity* 64 (2): 289–98.

Drummond, D. C. 1992. Unmasking Mascall's mouse traps. In *Proceedings of the Fifteenth Vertebrate Pest Conference 1992*, ed. J. E. Borrecco and R. E. Marsh, 229–35. Lincoln: University of Nebraska Press.

Dufraisse, A. 2006. Firewood economy during the 4th millennium B.C. at Lake Clairvaux, Jura, France. *Environmental Archaeology* 11 (1): 87–99.

Eastham, A. 1997. The potential of bird remains for environmental reconstruction. *International Journal of Osteoarchaeology* 7: 422–29.

Elder, W. H., and N. L. Elder. 1970. Social groupings and primate associations of the bushbucks (*Tragelaphus scriptus*). *Mammalia* 34: 356–62.

Ellis, E. C. 2011. Anthropogenic transformation of the terrestrial biosphere. *Philosophical Transactions of the Royal Society, London A* 369: 1010–35.

Elton, C. S. 1953. The use of cats in farm rat control. *British Journal of Animal Behaviour* 1 (4): 151–55.

Emery, N. 2005. Cognitive ornithology: The evolution of avian intelligence. *Proceedings of the Royal Society of London B* 361 (1465): 23–43.

Emery, N. J., and N. S. Clayton. 2004. The mentality of crows: Convergent evolution of intelligence in corvids and apes. *Science* 306 (5703): 1903–7.

Ericson, P., T. Tyrberg, A. S. Kjellberg, L. Jonsson, and I. Ullén. 1997. The earliest record of house sparrows (*Passer domesticus*) in northern Europe. *Journal of Archaeological Science* 24: 183–90.

Ervynck, A. 2002. Sedentism or urbanism? On the origin of the commensal black rat (*Rattus rattus*). In *Bones and the Man*, ed. K. M. Dobney and T. P. O'Connor, 95–109. Oxford: Oxbow Books.

Estabrook, A. H. 1907. The present status of the English Sparrow problem in America. *The Auk* 24: 129–34.

Evans, D. H. 2010. A good riddance of bad rubbish?: Scatological musings on rubbish disposal and the handling of "filth" in medieval and early post-medieval towns. In *Exchanging Medieval Material Culture: Studies in Archaeology and History Presented to Frans Verhaeghe, Relicta Monografieën 4*, ed. K. de Groote, T. Dries, and M. Pieters, 267–78. Brussels: Vlaams Instituut voor het Onroerend Erfgoed.

Evans, I. M., R. W. Summers, L. O'Toole, D. C. Orr-Ewing, R. Evans, N. Snell, and J. Smith. 1999. Evaluating the success of translocating red kites *Milvus milvus* to the UK. *Bird Study* 46 (2): 129–44.

Ezequiel, O. D. S., G. S. Gazeta, M. Amorim, and N. M. Serra-Freire. 2001. Evaluation of the acarofauna of the domiciliary ecosystem in Juiz de Fora, State of Minas Gerais, Brazil. *Memorias do Instituto Oswaldo Cruz* 96: 911–16.

Fabian, J. 1990. Presence and representation: The other and anthropological writing. *Critical Inquiry* 16 (4): 753–72.

Fahrig, L., and T. Rytwinski. 2009. Effects of roads on animal abundance: An empirical review and synthesis. *Ecology and Society* 14 (1): 1–20.

Fairnell, E. H., and J. H. Barrett. 2007. Fur-bearing species and Scottish islands. *Journal of Archaeological Science* 34: 463–84.

Feddersen-Petersen, D. U. 2008. Social behaviour of dogs and related canids. In *The Behavioural Biology of Dogs*, ed. P. Jensen, 105–19. Cambridge, MA: CABI International.

Finch, J. 2007. What more were the pastures of Leicester to me? Hunting, landscape character, and the politics of place. *International Journal of Cultural Property* 14: 361–83.

Fink, T. M. 2006. Of mice and men: What archaeologists should know about hantavirus and plague in North America. In *Dangerous Places: Health, Safety and Archaeology*, ed. D. A. Poirier and K. L. Feder, 31–53. Westport, CT: Bergin and Garvey.

Fiore, I., M. Gala, and A. Tagliacozzo. 2004. Ecology and subsistence strategies in the Eastern Italian Alps during the Middle Palaeolithic. *International Journal of Osteoarchaeology* 14: 273–86.

Fitzgerald, B. M. 1988. Diet of domestic cats and their impact on prey populations. In *The Domestic Cat: The Biology of Its Behaviour*, ed. D. C. Turner and P. Bateson, 123–215. Cambridge: Cambridge University Press.

Fitzwater, W. D. 1970. Trapping: The oldest profession. *Proceedings of the 4th Vertebrate Pest Conference 1970*, Sacramento, CA. Lincoln: University of Nebraska Press, 101–8.

Fladmark, K., J. C. Driver, and D. Alexander. 1988. The Paleoindian component at Charlie Lake Cave (HbRf 39), British Columbia. *American Antiquity* 53 (2): 371–84.

Flannery, T. F., and J. P. White. 1991. Animal translocation. *National Geographic Research and Exploration* 7 (1): 96–113.

Fleagle, J. G., J. J. Shea, F. E. Grine, A. L. Baden, and R. E. Leakey, eds. 2010. *Out of Africa I: The First Human Colonization of Eurasia.* New York: Springer.

Flinn, M. V. 1997. Culture and the evolution of social learning. *Evolution and Social Behaviour* 18: 23–67.

Fox, G. E. 1891. Excavations on the site of the Roman city at Silchester, Hants, in 1891: With a note on the animal remains found during the excavations by Herbert Jones Esq. *Archaeologia* 53: 263–88.

Frynta, D., M. Slábová, H. Váchová, R. Volfová, and P. Munclinger. 2005. Aggression and commensalism in house mouse: A comparative study across Europe and the Near East. *Aggressive Behaviour* 31: 283–93.

Fuentes, A. 2006. Human-nonhuman primate interconnections and their relevance to anthropology. *Ecological and Environmental Anthropology* 2 (2): 1–11.

Gade, D. W. 2010. Shifting synanthropy of the crow in eastern North America. *Geographical Review* 100 (2): 152–75.

Gadgill, M. 2001. Project Lifescape 9: Crows. *Resonance* (February): 74–82.

Ganem G., and J. B. Searle. 1996. Behavioural discrimination among chromosomal races of the house mouse (*Mus musculus domesticus*). *Journal of Evolutionary Biology* 9: 817–31.

García, G. O., M. Favero, and A. I. Vassallo. 2010. Factors affecting kleptoparasitism by gulls in a multi-species seabird colony. *The Condor* 112 (3): 521–29.

Garrod, D., and D. M. Bate. 1957. The Natufian culture: The life and economy of a Mesolithic people in the Near-East. *Proceedings of the British Academy* 43: 211–27.

Gehrt, S. D. 2007. Ecology of coyotes in urban landscapes. *Proceedings of the 12th Wildlife Damage Management Conference*, ed. D. L. Nolte, W. M. Arjo, and D. H. Stalman, 303–11. Lincoln: University of Nebraska Press.

Gehrt, S. D., J. L. Brown, and C. Anchor. 2011. Is the urban coyote a misanthropic synanthrope? The case from Chicago. *Cities and the Environment* 4 (1). http://digitalcommons.lmu.edu/cate/vol4/iss1/3/.

Gehrt, S. D., and S. Prange. 2006. Interference competition between coyotes and raccoons: A test of the mesopredator release hypothesis. *Behavioural Ecology* 18 (1): 204–14.

Gentry, A., J. Clutton-Brock, and C. P. Groves. 2004. The naming of wild animal species and their domestic derivatives. *Journal of Archaeological Science* 31: 645–51.

Gilbert, O. L., 1991. *The Ecology of Urban Habitats.* London: Chapman and Hall.

Gilbert-Norton, L. P., T. A. Shahan, and J. A. Shivak. 2009. Coyotes (*Canis latrans*) and the matching law. *Behavioural Processes* 82 (2): 178–83.

Gowlett, J. A. J. 2006. The early settlement of northern Europe: Fire history in the context of climate change and the social brain. *Comptes Rendus Palévolution* 5: 299–310.

Graham, J. H., and J. J. Duda. 2011. The humpbacked species richness-curve: A contingent rule for community ecology. *International Journal of Ecology* 2011: 1–15.

Gray, M. M., N. B. Sutter, E. A. Ostrander, and R. K. Wayne. 2010. The IGF1 small dog haplotype is derived from Middle Eastern grey wolves. *BMC Biology* 8: 1–13.

Grayson, D. K. 2001. The archaeological record of human impact on animal populations. *Journal of World Prehistory* 15 (1): 1–68.

Grine, F. E., J. G. Fleagle, and R. E. Leakey, eds. 2009. *The First Humans: Origin and Early Evolution of the Genus Homo.* New York: Springer Science and Business.

Gullone, E. 2000. The Biophilia Hypothesis and life in the 21st century: Increasing mental health or increasing pathology? *Journal of Happiness Studies* 1: 293–321.

Gyllensten, U., and A. C. Wilson. 1987. Interspecific mitochondrial DNA transfer and the colonisation of Sweden by mice. *Genetical Research* 49: 25–29.

Haag-Wackernagel, D. 2006. Human diseases caused by feral pigeons. *Advances in Vertebrate Pest Management* 4: 31–58.

Haag-Wackernagel, D., and I. Geigenfand. 2008. Protecting buildings against feral pigeons. *European Journal of Wildlife Research* 54: 715–21.

Haag-Wackernagel, D., and H. Moch. 2004. Health hazards posed by feral pigeons. *Journal of Infection* 48: 307–13.

Haberle, S. G. 2007. Prehistoric human impact on rainforest diversity in highland New Guinea. *Philosophical Transactions of the Royal Society of London B* 362: 219–28.

Haensch, S., R. Bianucci, M. Signoli, M. Rajerison, M. Schultz, S. Kacki, M. Vermunt, D. A. Weston, D. Hurst, M. Achtman, E. Carniel, and B. Bramanti. 2010. Distinct clones of *Yersinia pestis* caused the Black Death. *PLoS Pathogens* 6 (10): e1001134. doi:10.1371/journal.ppat.1001134.

Hamilakis, Y. 2001. Re-inventing environmental archaeology. In *Environmental Archaeology: Meaning and purpose*, ed. U. Albarella, 29–38. Dordrecht, Netherlands: Kluwer Scientific.

Harcourt, R. 1979. The animal bones. In *Gussage All Saints: An Iron Age settlement in Dorset*, ed. G. J. Wainwright, 150. London: HMSO, DoE Archaeological Reports 10.

Harcourt-Smith, W. H. E. 2010. The first hominins and the origins of bipedalism. *Evolution: Education and Outreach* 3 (3): 333–40.

Harper, G. A. 2005. Numerical and functional response of feral cats (*Felis catus*) to variations in abundance of primary prey on Stewart Island (Rakiura), New Zealand. *Wildlife Research* 32: 597–604.

Harris, S. 1977. Distribution, habitat utilization and age structure of a suburban fox (*Vulpes vulpes*) population. *Mammal Review* 7 (1): 25–38.

Harris, S., and J. M. V. Rayner. 1986. Urban fox (*Vulpes vulpes*) population estimates and habitat requirements in several British cities. *Journal of Animal Ecology* 55: 575–91.

Harrison, S. W. R., B. L. Cypher, S. Bremner-Harrison, and C. Van Horn Job. 2011. Resource overlap between urban carnivores: Implications for endangered San Joaquin kit foxes (*Vulpes macrotis mutica*). *Urban Ecosystems* 14 (2): 303–11.

Harrison, N. M., and M. J. Whitehouse. 2011. Mixed-species flocks: An example of niche construction? *Animal Behaviour* 81 (4): 675–82.

Hart, D. L., and R. W. Sussman. 2009. *Man the Hunted.* Boulder, CO: Westview Press.

Haslam, M., A. Hernandez-Aguilar, V. Ling, S. Carvalho, I. de la Torre, A. DeStefano, A. Du, B. Hardy, J. Harris, L. Marchant, T. Matsuzawa, W. McGrew, J. Mercader, R. Mora, M. Petraglia, H. Roche, E. Visalberghi, and R. Warren. 2009. Primate archaeology. *Nature* 460: 339–44.

Heinsohn, T. E. 2001. Human influences on vertebrate zoogeography: Animal translocation and biological invasions across and to the east of Wallace's Line. In *Faunal and Floral Migrations and Evolution in SE Asia–Australasia*, ed. I. Metcalfe, J. M. B. Smith, M. Morwood, and I. Davidson, 153–70. Lisse, Netherlands: Swets & Zeitlinger.

Heinsohn, T. E. 2003. Animal translocation: Long-term human influences on the vertebrate zoogeography of Australasia (natural dispersal versus ethnophoresy). *Australian Zoologist* 32 (3): 351–76.

Herr, J., L. Schley, and T. J. Roper. 2009. Socio-spatial organisation of urban stone martens. *Journal of Zoology* 277 (1): 54–62.

Herzog, H. 2002. Darwinism and the study of human-animal interactions. *Society and Animals* 10 (4): 361–67.

Hinton, M. A. C. 1930. *Rats and Mice as Enemies of Mankind.* 3rd ed. London: British Museum of Natural History.

Hole, F. 1984. A reassessment of the Neolithic Revolution. *Paléorient* 10 (2): 49–60.

Honeycutt, R. L. 2010. Unraveling the mysteries of dog evolution. *BMC Biology* 8: 20. http://www.biomedcentral.com/1741-7007/8/20.

Horsburgh, K. 2008. Wild or domesticated? An ancient DNA approach to canid species identification in South Africa's Western Cape Province. *Journal of Archaeological Science* 35 (6): 1474–80.

Horton, M. C. 1996. *Shanga: A Muslim Trading Community on the East African Coast.* Nairobi, Kenya: British Institute in Eastern Africa.

Hounslow, M., T. Johnson, A. Kathan, and H. Pound. 2010. *Animal Abuse and Empathy in Children.* Mount Royal University and Calgary Humane Society. https://mtroyal.ca/wcm/groups/public/documents/pdf/chs_finalreport_pdf.pdf (consulted 15 February 2012).

Hughes, J. K., S. Elton, and H. J. O'Regan. 2008. Theropithecus and "Out of Africa" dispersal in the Plio-Pleistocene. *Journal of Human Evolution* 54: 43–77.

Husselman, E. M. 1953. The dovecotes of Karanis. *Transactions and Proceedings of the American Philological Association* 84: 81–91.

Hutchinson, G. E. 1978. *An Introduction to Popular Ecology.* New Haven, CT: Yale University Press.

İmamoğlu, V., M. Korumaz, and C. İmamoğlu. 2005. A fantasy in Central Anatolian architectural heritage: Dovecotes and towers in Kayserí. *METU JFA* 22 (2): 79–90.

Indykiewicz, P. 2005. Factors determining number fluctuations and variation of the breeding success of an urban population of the Black-headed Gull *Larus ridibundus* (N-Poland). *Folia biologica (Kraków)* 53: 165–69.

Irwin, P., C. Arcari, J. Hausbeck, and S. Paskowitz. 2006. Urban wet environment as mosquito habitat in the Upper Midwest. *Ecohealth* 5 (1): 49–57.

Isacoff, J. 2005. Writing the Arab-Israeli conflict: Historical bias and the use of history in political science. *Perspectives on Politics* 3: 71–88.

Jacquin, L., B. Cazelles, A-C. Prévot-Julliard, G. Leboucher, and J. Gasparini. 2010. Reproduction management affects breeding ecology and reproduction costs in feral urban pigeons (*Columba livia*). *Canadian Journal of Zoology* 88: 781–87.

James, S. R. 1989. Hominid use of fire in the Lower and Middle Pleistocene. *Current Anthropology* 30: 1–26.

Jankowiak, L., M. Antczak, and P. Tryjanowski. 2008. Habitat use, food, and the importance of chicken in the diet of the red fox (*Vulpes vulpes*) on extensive farmland in Poland. *World Applied Sciences Journal* 4 (6): 886–90.

Jerolmack, C. 2008. How pigeons became rats: The cultural-spatial logic of problem animals. *Social Problems* 55 (1): 72–94.

Johnston, R. F. 2001. Synanthropic birds of North America. In *Avian Ecology in an Urbanizing World,* ed. J. M. Marzluff, R. Bowman, and R. Donnelly, 49–67. Norwell: Kluwer Academic Publishers.

Jones, C. G., J. H. Lawton, and M. Shachak. 1994. Organisms as ecosystem engineers. *Oikos* 69: 373–86.

Jones, D. N., and S. J. Reynolds. 2008. Feeding birds in our towns and cities: A global research opportunity. *Journal of Avian Biology* 39 (3): 265–71.

Joye, Y., and A. de Block. 2011. "Nature and I are two": A critical examination of the Biophilia Hypothesis. *Environmental Values* 20: 189–215.

Juma, A. 2004. *Unguja Ukuu on Zanzibar: An Archaeological Study of Early Urbanism.* Uppsala, Sweden: Societas Archaeologica Upsaliensis.

Kahn, P. H., R. L. Severson, and J. H. Ruckert. 2009. The human relation with nature and technological nature. *Current Directions in Psychological Science* 18 (1): 37–42.

Kajiura, S. M., L. J. Macesic, T. L. Meredith, K. L. Cocks, and L. J. Dirk. 2009. Commensal foraging between double-crested cormorants and a southern stingray. *Wilson Journal of Ornithology* 121 (3): 646–48.

Kalof, L. 2007. *Looking at Animals in Human History*. London: Reaktion Books.

Karavanić, I., P. T. Miracle, M. Culiberg, D. Izurtanek, J. Zupanič, V. Golubić, M. Paunović, J. Mauch, V. Malez, R. Šošić, I. Janković, and F. H. Smith. 2008. The Middle Palaeolithic from Mujina Pećina, Dalmatia, Croatia. *Journal of Field Archaeology* 33: 259–77.

Karkanas, P., R. Shahack-Gross, A. Ayalon, M. Bar-Matthews, R. Barkai, A. Frumkin, A. Gopher, and M. C. Stiner. 2007. Evidence for habitual use of fire at the end of the Lower Palaeolithic: Site formation processes at Qesem Cave, Israel. *Journal of Human Evolution* 53: 197–212.

Keene, D. J. 1981. Rubbish in medieval towns. In *Environmental Archaeology in the Urban Context*, ed. R. A. Hall and H. K. Kenward, 26–30. CBA Research Report 43. London: Council for British Archaeology.

Kellert, S. R., and Wilson, E. O., eds. 1993. *The Biophilia Hypothesis*. Washington, DC: Island Press.

Kimura, B., F. B. Marshall, S. Chen, S. Rosenbom, P. D. Moehlman, N. Tuross, R. C. Sabin, J. Peters, B. Barich, H. Yohannes, F. Kebede, R. Teclai, A. Beja-Pereira, and C. Mulligan. 2010. Ancient DNA from Nubian and Somali wild ass provides insights into donkey ancestry and domestication. *Proceedings of the Royal Society B* 278 (1702): 50–57.

King, A. 2005. Animal remains from temples in Roman Britain. *Britannia* 36: 329–69.

King, A. J., and G. Cowlishaw. 2009. Foraging opportunities drive interspecific associations between rock kestrels and desert baboons. *Journal of Zoology* 277 (2): 111–18.

Kitchener, A. C., and T. P. O'Connor. 2010. Wildcats, domestic and feral cats. In *Extinctions and Invasions: A Social History of British Fauna*, ed. T. P. O'Connor and N. Sykes, 83–94. Oxford: Windgather Press.

Knight, J. 1999. Monkeys on the move: The natural symbolism of people-macaque conflict in Japan. *Journal of Asian Studies* 58 (3): 622–47.

Kociolek, A. V., A. P. Clevenger, C. C. St Clair, and D. S. Proppe. 2010. Effects of road networks on bird populations. *Conservation Biology* 25 (2): 241–49.

König, A. 2008. Fears, attitudes and opinions of suburban residents with regards to their urban foxes. *European Journal of Wildlife Research* 54: 101–9.

Kruuk, H. 1972. Surplus killing by carnivores. *Journal of Zoology* 166 (2): 233–44.

Landon, D. B. 2009. An update on zooarchaeology and historical archaeology: Progress and prospects. In *International Handbook of Historical Archaeology*, ed. T. Majewski and D. Gaimster, 77–104. New York: Springer.

Langton, S. D., D. P. Cowan, and A. N. Meyer. 2001. The occurrence of commensal rodents in dwellings as revealed by the 1996 English House Condition Survey. *Journal of Applied Ecology* 38 (4): 699–709.

Leach, M. C., and D. C. Main. 2008. An assessment of laboratory mouse welfare in UK animal units. *Animal Welfare* 17 (2): 171–87.

Lee, P., and N. Priston. 2005. Human attitudes to primates: Perceptions of pests, conflict and consequences for primate conservation. In *Commensalism and Conflict: The Primate-Human Interface*, ed. J. D. Patterson, 1–23. Winnipeg, MB: Hignell Printing.

Lesnik, J., and J. F. Thackeray. 2007. The efficiency of stone and bone tools for opening termite mounds: Implications for hominin tool use at Swartkrans. *South African Journal of Science* 103 (9–10): 354–56.

Le Souef, J. C. 1958. The introduction of sparrows into Victoria. *Emu* 12 (3): 190.

Lever, C. 2009. *The Naturalized Animals of Britain and Ireland.* New York: New Holland Publishers.

Levy, S. S. 2003. The Biophilia Hypothesis and anthropocentric environmentalism. *Environmental Ethics* 25: 227–46.

Liberg, O., and M. Sandell. 1988. Spatial organisation and reproductive tactics in the domestic cat and other felids. In *The Domestic Cat: The Biology of Its Behaviour*, ed. D. C. Turner and P. Bateson, 159–77. Cambridge: Cambridge University Press.

Long, J. L. 1981. *Introduced Birds of the World.* Newton Abbott, UK: David and Charles.

Long, J. L. 2003. *Introduced Mammals of the World: Their History, Distribution and Influence.* Collingwood: CABI.

Lord, T. C., T. P. O'Connor, D. Siebrandt, and R. Jacobi. 2007. People and large carnivores as biostratinomic agents in Lateglacial cave assemblages. *Journal of Quaternary Science* 22 (7): 681–94.

Losos, J. R. 2000. Ecological character displacement and the study of adaptation. *Proceedings of the National Academy of Sciences* 97 (11): 5693–95.

Louv, R. 2008. *Last Child in the Woods: Saving Our Children from Nature Deficit Disorder.* New York: Workman Publishing Co.

Lowe, B. 2005. The domestic cat in Egyptian tomb paintings. *The Ostracon* 16 (1): 7–11.

Lundholm, J. T., and P. J. Richardson. 2010. Habitat analogues for reconciliation ecology in urban and industrial environments. *Journal of Applied Ecology* 47: 966–75.

Luniak, M. 2004. Synurbization—Adaptation of animal wildlife to urban development. In *Urban Wildlife Conservation. Proceedings 4th International Urban Wildlife Symposium*, ed. W. W. Shaw, K. L. Harris, and L. Van Druff, 50–55. Tucson: University of Arizona Press.

Lukasz, R. 2001. Feeding activity and seasonal change in prey consumption of urban peregrine falcons *Falco peregrinus. Acta Ornithologica* 36: 165–69.

Lyman, R. L. 1994. *Vertebrate Taphonomy.* Cambridge: Cambridge University Press.

Macholán, M. 2006. A geometric morphometric analysis of the shape of the first upper molar in mice of the genus *Mus* (Muridae, Rodentia). *Journal of Zoology* 270 (4): 672–81. doi:10.1111/j.1469-7998.2006.00156.x.

Maciusik, B., M. Lenda, and P. Skórka. 2010. Corridors, local food resources, and climatic conditions affect the utilization of the urban environment by Black-Headed Gull *Larus ridibundus* in winter. *Ecological Research* 25: 263–72.

Maher, L. A., J. T. Stock, S. Finney, J. J. N. Heywood, P. Miracle, and E. B. Banning. 2011. A unique human-fox burial from a pre-Natufian cemetery in the Levant (Jordan). *PLoS ONE* 6 (1): e15815.

Majewska, A. C., T. K. Graczyk, A. Słodkowicz-Kowalska, L. Tamang, S. Jedrzewski, P. Zduniak, P. Solarczyk, A. Nowosad, and P. Nowosad. 2009. The role of free-ranging, captive, and domestic birds of Western Poland in environmental contamination with *Cryptosporidium parvum* oocysts and *Giardia lamblia* cysts. *Parasitology Research* 104: 1093–99.

Majolo, B., and R. Ventura. 2004. Apparent feeding association between Japanese macaques (*Macaca fuscata yakui*) and sika deer (*Cervus nippon*) living on Yakushima Island, Japan. *Ethology, Ecology and Evolution* 16: 33–40.

Makarewicz, C. A. 2009. Complex caprine harvesting practices and diversified hunting strategies: Integrated animal exploitation systems at Late Pre-Pottery Neolithic A 'Ain Jamman. *Anthropozoologica* 44 (1): 79–101.

Malek, J. 1993. *The Cat in Ancient Egypt.* London: British Museum Press.

Maller, C. J., C. Henderson-Wilson, and M. Townsend. 2010. Rediscovering nature in everyday settings: Or how to create healthy environments and healthy people. *EcoHealth* 6 (4): 553–56. doi: 10.1007/s10393-010-0282-5.

Maltby, M. 2010. *Feeding a Town: Environmental Evidence from Excavations in Winchester, 1972–1985.* Winchester, UK: Winchester Museum Service.

Manning, A., and J. Serpell. 1994. *Animals and Human Society: Changing Perspectives.* London: Routledge.

Mannu, M., and E. Ottoni. 2009. The enhanced tool-kit of two groups of wild bearded capuchin monkeys in the Caatinga: Tool making, associative use, and secondary tools. *American Journal of Primatology* 71: 242–51.

Marshall, J. T., Jr., and R. D. Sage. 1981. Taxonomy of the house mouse. *Symposia of the Zoological Society of London* 47: 15–25.

Martin, L., and L. Fitzgerald. 2005. A taste for novelty in invading house sparrows *Passer domesticus. Behavioural Ecology* 16: 702–7.

Martin, R. E. 1999. *Taphonomy: A Process Approach.* Cambridge: Cambridge University Press.

Marzluff, J., and T. Angell. 2005. Cultural co-evolution: How the human bond with crows and ravens extends theory and raises new questions. *Journal of Ecological Anthropology* 9: 69–75.

Marzluff, J., E. Schulenberger, W. Endlicher, M. Alberti, G. Bradley, C. Ryan, C. ZumBrunnen, and U. Simon, eds. 2008. *Urban Ecology: An International Perspective on the Interaction between Humans and Nature.* New York: Springer Science and Business Media.

Mashkour, M., H. Monchot, E. Trinkaus, J.-L. Reyss, F. Biglan, S. Bailon, S. Heydan, and K. Abdi. 2009. Carnivores and their prey in the Wezmeh Cave (Kermanshah, Iran): A Late Pleistocene refuge in the Zagros. *International Journal of Osteoarchaeology* 19 (6): 678–94.

Matisoo-Smith, E., and J. H. Robins. 2004. Origins and dispersals of Pacific peoples: Evidence from mtDNA phylogenies of Pacific rat. *Proceedings of the National Academy of Sciences* 101 (24): 9167–72.

Matisoo-Smith, E., and J. Robins. 2009. Mitochondrial DNA evidence for the spread of Pacific rats through Oceania. *Biological Invasions* 11: 1521–27.

Mattingley, D. 2007. *An Imperial Possession: Britain in the Roman Empire, 54 BC–AD 409.* London: Penguin Books.

McClure, B. 2011. Hedgehogs: Basic care and first aid. *Veterinary Nursing Journal* 26 (7): 238–40.

McGrew, W. C. 2004. *The Cultured Chimpanzee: Reflections on Cultural Primatology.* Cambridge: Cambridge University Press.

McGrew, W. C. 2010. Chimpanzee technology. *Science* 328 (5978): 579–80.

McGuire, B., T. Pizzuto, W. E. Bemiss, and L. L. Getz. 2006. General ecology of a rural population of Norway rats (*Rattus norvegicus*) based on intensive live trapping. *American Midland Naturalist* 155 (1): 221–36.

McKinney, M. L. 2006. Urbanization as a major cause of biotic homogenization. *Biological Conservation* 127: 247–60.

McKinney, T. 2010. The effect of provisioning and crop-raiding on the diet and foraging activities of human-commensal white-faced capuchins (*Cebus capucinus*). *American Journal of Primatology* 71 (1): 1–10.

Meadow, R. G. 1989. Osteological evidence for the process of animal domestication. In *The Walking Larder*, ed. J. Clutton-Brock, 80–90. Boston: Unwin Hyman.

Mighall, T., S. Timberlake, I. D. L. Foster, E. Krupp, and S. Singh. 2009. Ancient copper and lead pollution records from a raised bog complex in Central Wales, UK. *Journal of Archaeological Science* 36 (7): 1504–15.

Miller, J. R. 2008. Conserving biodiversity in metropolitan landscapes: A matter of scale (but which scale?). *Landscape Journal* 27: 114–26.

Milne, C., and P. Crabtree. 2001. Prostitutes, a rabbi and a carpenter: Dinner at the Five Points in the 1830s. *Historical Archaeology* 35 (3): 31–48.

Mithen, S. 1999. The hunter-gatherer prehistory of human-animal interactions. *Anthrozoos* 12: 195–204.

Miyata, H., and K. Fujita. 2008. Pigeons (*Columba livia*) plan future moves on computerized maze tasks. *Animal Cognition* 11: 505–16.

Moore, A. M. T., G. C. Hillman, and A. J. Legge. 2000. *Village on the Euphrates: From Foraging to Farming at Abu Hureyra*. New York: Oxford University Press.

Morales Muñiz, A., M. A. Pecharroman, F. H. Carrasquilla, and C. Liesau von Lettow-Vorbeck. 1995. Of mice and sparrows: Commensal faunas from the Iberian Iron Age in the Duero Valleu (Central Spain). *International Journal of Osteoarchaeology* 5: 127–38.

Münzel, S., and N. Conard. 2004. Change and continuity in subsistence during the Middle and Upper Palaeolithic. *International Journal of Osteoarchaeology* 14: 225–43.

Murton, R. K., R. J. P. Thearle, J. Thompson. 1972a. Ecological studies of the feral pigeon Columba livia var. I. Population, breeding biology and methods of control. *Journal of Applied Ecology* 9 (3): 835–74.

Murton, R. K., C. F. B. Coombs, and R. J. P. Thearle. 1972b. Ecological studies of the feral pigeon Columba livia var. II. Flock behaviour and social organization. *Journal of Applied Ecology* 9 (3): 875–89.

Nadel, D., G. Lengyel, F. Bocquetin, A. Tsatskin, D. Rosenberg, R. Yeshurun, G. Bar-Oz, D. Bar-Yosek Mayer, R. Beeri, L. Conyers, I. Hershkovitz, A. Kurowska, and L. Weissbrod. 2008. The late Natufian at Reqefet Cave: The 2006 excavation season. *Journal of the Israel Prehistoric Society* 38: 59–131.

Natoli, E., and E. De Vito. 1991. Agonistic behaviour, dominance rank and copulatory success in a large multi-male feral cat (*Felis catus* L.) colony in central Rome. *Animal Behaviour* 42 (2): 227–41.

Naughton-Treves, L., A. Treves, A. C. Chapman, and R. Wrangham. 1998. Temporal patterns of crop-raiding by primates: Linking food availability in croplands and adjacent forest. *Journal of Applied Ecology* 35: 596–606.

Nicastro, N. 2004. Perceptual and acoustic evidence for species-level differences in meow vocalizations by domestic cats (*Felis catus*) and African wild cats (*Felis silvestris lybica*). *Journal of Comparative Psychology* 118: 287–96.

Niemelä, J. 1999. Ecology and urban planning. *Biodiversity and Conservation* 8: 119–31.

Nihei, Y., and H. Higuchi. 2001. When and where did crows learn to use automobiles as nutcrackers? *Tohoku Psychologica Folia* 60: 93–97.

Nogales, M., A. Martín, B. R. Tershy, C. J. Donlan, B. Veitch, N. Perto, B. Wood, and J. Alonso. 2004. A review of feral cat eradication on islands. *Conservation Biology* 18 (2): 310–19.

Novákova, M., R. Palme, H. Kutalová, L. Jansky, and D. Frynta. 2008. The effect of sex, age and commensal way of life on levels of fecal glucocorticoid metabolites in spiny mice (*Acomys cahirinus*). *Physiology and Behaviour* 95: 187–213.

O'Connor, T. P. 1986. The animal bones. In *The Legionary Fortress Baths at Caerleon*, vol. 2, *The Finds*, ed. D. Zienkiewicz, 223–46. Cardiff: National Museum of Wales.

O'Connor, T. P. 1991. *Bones from 46–54 Fishergate*. Archaeology of York 15/4. London: Council for British Archaeology.

O'Connor, T. P. 1992. Pets and pests in Roman and medieval Britain. *Mammal Review* 22 (2): 107–13.

O'Connor, T. P. 1997. Working at relationships: Another look at animal domestication. *Antiquity* 71 (271): 149–56.

O'Connor, T. P. 2000a. *The Archaeology of Animal Bones*. College Station: Texas A&M University Press.

O'Connor, T. P. 2000b. Human refuse as a major ecological factor in medieval urban vertebrate communities. In *Human Ecodynamics*, ed. G. Bailey, R. Charles, and N. Winder, 15–20. Oxford: Oxbow Books.

O'Connor, T. P. 2007. Wild or domestic? Biometric variation in the cat *Felis silvestris* Schreber. *International Journal of Osteoarchaeology* 17 (6): 581–95. doi: 10.1002/oa.913.

O'Connor, T. P. 2010a. Making themselves at home: The archaeology of commensal vertebrates. In *Anthropological Approaches to Zooarchaeology*, ed. D. Campana, P. Crabtree, S. D. de France, J. Lev-Tov, and A. Choyke, 270–74. Oxford: Oxbow Books.

O'Connor, T. P. 2010b. Culture and environment: Mind the gap. In *Land and People: Papers in memory of John G. Evans*, ed. M. J. Allen, N. Sharples, and T. P. O'Connor, 11–18. Prehistoric Society Research Paper 2. London: The Prehistoric Society.

O'Connor, T. P. 2010c. The house mouse. In *Extinctions and Invasions: A Social History of British Fauna*, ed. T. P. O'Connor and N. Sykes, 127–33. Oxford: Windgather Press.

O'Connor, T. P. 2011. The changing environment of the Yorkshire Dales during the Late and Post-Glacial periods. In *The Prehistory of the Yorkshire Dales*, ed. R. Martlew. York: People, Landscape and Cultural Environment of Yorkshire. http://place.uk.com/2011/07/publication-of-prehistory-in-the-yorkshire-dales/.

O'Connor, T. P., and J. G. Evans. 2005. *Environmental Archaeology: Principles and Methods.* 2nd ed. Stroud: Sutton Publishing.

Okubo, A., P. K. Maini, M. H. Williamson, and J. D. Murray. 1989. On the spatial spread of the grey squirrel in Britain. *Proceedings of the Royal Society of London B* 283: 113–25.

Osłowksi, G. 2008. Roadside hedgerows and trees as factors increasing road mortality of birds: Implications for management of roadside vegetation in rural landscapes. *Biological Conservation* 86: 153–61.

Otali, E., and J. S. Gilchrist. 2004. The effects of refuse feeding on body condition, reproduction and survival of banded mongooses. *Journal of Mammalogy* 85 (3): 491–97.

Ottoni, E. B., and P. Izar. 2008. Capuchin monkey tool use: Overview and implications. *Evolutionary Anthropology* 17 (4): 171–78.

Parker, A. J. 1988. Birds of Roman Britain. *Oxford Journal of Archaeology* 7 (2): 197–226.

Parslow, J. L. F. 1967. Changes in status among breeding birds in Britain and Ireland. *British Birds* 60: 177–202.

Paulos, E., R. J. Honicky, B. Hooker. 2008. Citizen science: Enabling participatory urbanism. In *Handbook of Research on Urban Informatics: The Practice and Promise of the Real-Time City*, ed. M. Foth, 1–16. Hershey, PA: Information Science Reference, IGI Global.

Pavao-Zuckerman, B. 2007. Deerskins and domesticates: Creek subsistence and economic strategies in the historic period. *American Antiquity* 72 (1): 5–33.

Pearre, S., and R. Maass. 1998. Trends in the prey size–based trophic niches of feral and House Cats *Felis catus* L. *Mammal Review* 28 (3): 125–39.

Peresani, M., I. Fiore, M. Gala, M. Romandini, and A. Tagliacozo. 2011. Late Neandertals and the intentional removal of feathers as evidenced from bird bone taphonomy at Fumane Cave 44 ky B.P., Italy. *Proceedings of the National Academy of Sciences* 108 (10): 3888–93.

Peters, J., and K. Schmidt. 2004. Animals in the symbolic world of Pre-Pottery Neolithic Göbekli Tepe, south-eastern Turkey: Preliminary assessment. *Anthropozoologica* 39 (1): 179–218.

Pickford, M. 2005. Orientation of the foramen magnum in Late Miocene to extant African apes and hominids. *Anthropologie* 43 (2–3): 103–10.

Pigière, F., and D. Henrotay. 2011. Camels in the northern provinces of the Roman Empire. *Journal of Archaeological Science* 39 (5): 1531–39. http://dx.doi.org/10.1016/j.jas.2011.11.014.

Piper, P. J., F. Z. Campos, D. Ngoc Kinh, N. Amano, M. Oxenham, B. Chi Hoang, P. Bellwood, and A. Willis. 2012. Early evidence for pig and dog husbandry from the Neolithic site of An Son, Southern Vietnam. *International Journal of Osteoarchaeology.* doi: 10.1002/oa.2226; published online 6 February 2012.

Piperno, D., and T. D. Dillehay. 2008. Starch grains on human teeth reveal early broad crop diet in northern Peru. *Proceedings of the National Academy of Science* 105 (50): 19622–27.

Plug, I. 1993. The macrofaunal remains of wild animals from Abbot's Cave and Lame Sheep Shelter, Seacow Valley, Cape. *Koedoe* 36 (1): 15–26.

Pobiner, B. L., M. J. Rogers, C. M. Monahan, and J. W. K. Harris. 2008. New evidence for carcass processing strategies at 1.5Ma, Koobi Fora, Kenya. *Journal of Human Evolution* 55 (1): 103–30.

Podzorski, P. V. 1990. *Their Bones Shall Not Perish: An Examination of Predynastic Human Skeletal Remains from Naga-ed-Dêr in Egypt.* New Malden, UK: Sia Publishing.

Polis, G. A., and D. R. Strong. 1996. Food web complexity and community dynamics. *American Naturalist* 147 (5): 813–46.

Potts, G. R. 1967. Urban starling roosts in the British Isles. *Bird Study* 14 (1): 25–42.

Prager, E. M., R. D. Sage, U. Gyllensten, W. Kelley Thomas, R. Rubner, C. S. Jones, L. Noble, J. B. Searle, and A. C. Wilson. 1992. Mitochondrial DNA sequence diversity and the colonization of Scandinavia by house mice from East Holstein. *Biological Journal of the Linnaean Society* 50 (2): 85–122.

Prange, S., and S. D. Gehrt. 2004. Changes in mesopredator community structure in response to urbanization. *Canadian Journal of Zoology* 82 (11): 1804–17.

Prange, S., and S. D. Gehrt. 2007. Response of skunks to a simulated increase in coyote activity. *Journal of Mammalogy* 88 (4): 1040–49.

Prange, S., S. D. Gehrt, and E. P. Wiggers. 2003. Demographic factors contributing to high raccoon densities in urban landscapes. *Journal of Wildlife Management* 67 (2): 324–33.

Prange, S., S. D. Gehrt, and E. P. Wiggers. 2004. Influence of anthropogenic resources in raccoon (*Procyon lotor*) movements and spatial distribution. *Journal of Mammalogy* 85 (3): 483–90.

Preece, R. C., J. A. J. Gowlett, S. A. Parfitt, D. R. Bridgland, and S. G. Lewis. 2006. Humans in the Hoxnian: Habitat, context and fire use at Beeches Pit, West Stow, Suffolk UK. *Journal of Quaternary Science* 21 (5): 485–96.

Pruvost, M., R. Bellone, N. Benecke, E. Sandoval-Castellanos, M. Cieslaka, T. Kuznetsova, A. Morales-Muñiz, T. O'Connor, M. Reissmann, M. Hofreiter, and A. Ludwig. 2011. Genotypes of pre-domestic horses match phenotypes painted in Paleolithic works of cave art. *Proceedings of the National Academy of Sciences of America* 108 (46): 18626–30.

Pyle, R. M. 1993. *The Thunder Tree: Lessons from an Urban Wildland.* Boston: Houghton Mifflin.

Ramsay, J., and Y. Tepper. 2010. Signs from a green desert: A preliminary examination of the archaeobotanical remains from a Byzantine dovecote site near Shivta, Israel. *Vegetation History and Archaeobotany* 19: 235–42.

Ranere, A. J., D. R. Piperno, I. Holst, R. Dickan, and J. Iriarte. 2009. The cultural and chronological context of early Holocene maize and squash domestication in the Central Bahas River Valley, Mexico. *Proceedings of the National Academy of Science* 106 (13): 5014–18.

Rathje, W. L., and C. Murphy. 1992. *Rubbish! The Archaeology of Garbage.* New York: Harper Collins.

Rathje, W. L., W. W. Hughes, D. C. Wilson, M. K. Tani, G. H. Archer, R. G. Hunt, and T. W. Jones. 1992. The archaeology of contemporary landfills. *American Antiquity* 57 (3): 437–47.

Réale, D., S. M. Reader, D. Sol, P. T. McDougall, and M. J. Dingemanse. 2007. Integrating animal temperament within ecology and evolution. *Biology Reviews* 82: 291–318.

Rebele, F. 1994. Urban ecology and special features of urban ecosystems. *Global Ecology and Biogeography Letters* 4: 173–87.

Reinhard, K. J., and V. M. Bryant. 1992. Coprolite analysis: A biological perspective on archaeology. *Journal of Archaeological Method and Theory* 4: 245–88.

Reitz, E., and N. Honerkamp. 1983. British colonial subsistence strategy on the Southeastern Coastal Plain. *Historical Archaeology* 17 (2): 4–26.

Reitz, E., and E. S. Wing. 2006. *Zooarchaeology*. 2nd ed. Cambridge: Cambridge University Press.

Rice, S., L. Brown, and H. S. Caldwell. 1973. Animals and psychotherapy: A survey. *Journal of Community Psychology* 1: 323–26.

Richmond, B. G., and W. L. Jungers. 2008. *Orrorin tugenensis* femoral morphology and the evolution of hominin bipedalism. *Science* 319 (5870): 1662–65.

Richter, T., A. N. Garrard, S. Allcock, and L. A. Maher. 2011. Interaction before agriculture: Exchanging material and sharing knowledge in the Final Pleistocene Levant. *Cambridge Archaeological Journal* 21: 95–114.

Rielly, K. 2010. The Black Rat. In *Extinctions and Invasions. A Social History of British Fauna*, ed. T. O'Connor and N. Sykes, 134–45. Oxford: Windgather Press.

Robardet, E., P. Girardoux, C. Caillot, D. Auget, F. Boue, and J. Barrat. 2011. Fox defecation behaviour in relation to spatial distribution of voles in an urbanised area: An increasing risk of transmission of *Echinococcus multilocularis*? *International Journal for Parasitology* 41 (2): 145–54.

Rock, P. 2005. Urban gulls: Problems and solutions. *British Birds* 98: 335–88.

Rollefson, G. O. 2008. Charming lives: Human and animal figurines in the Late Epipaleolithic and Early Neolithic periods in the Greater Levant and Eastern Anatolia. In *The Neolithic Demographic Transition and Its Consequences*, ed. J. P. Bocquet-Appel and O. Bar-Yosef, 387–416. New York: Springer Science and Business.

Rose, E., P. Nagel, and D. Haag-Wackernagel. 2006. Spatio-temporal use of the urban habitat by feral pigeons (*Columba livia*). *Behavioural Ecology and Sociobiology* 60: 242–54.

Roxburgh, S. H., K. Shea, and J. B. Wilson. 2004. The intermediate disturbance hypothesis: Patch dynamics and mechanisms of species co-existence. *Ecology* 85 (2): 359–71.

Ruddiman, W. F. 2007. The early anthropogenic hypothesis: Challenges and responses. *Reviews of Geophysics* 45 (RG4001): 1–37. doi: 10.1029/2006.RG000207.

Rutz, C. 2008. The establishment of an urban bird population. *Journal of Animal Ecology* 77 (5): 1008–19. doi: 10.1111/j.1365-2656.2008.01420.x.

Ruiz, M. O., E. D. Walker, E. S. Foster, L. D. Haramis, and U. D. Kitron. 2007. Association of West Nile virus and urban landscape in Chicago and Detroit. *International Journal of Health Geographics* 6 (10). doi: 10.1186/1476-072X-6-10. http://ij-healthgeographics.com/content/6/1/10 (accessed 21 October 2011).

Russell, N. 2002. The wild side of animal domestication. *Society and Animals* 10 (3): 285–302.

Saña Segui, M., D. Helmer, J. Peters, and A. von den Driesch. 1999. Early animal husbandry in the northern Levant. *Paléorient* 25 (2): 27–48.

Sapir-Hen, L., G. Bar-Oz, H. Kalaily, and T. Dayan. 2009. Gazelle exploitation at the early Neolithic site of Motza, Israel: The last of the gazelle hunters in the southern Levant. *Journal of Archaeological Science* 36: 1538–46.

Sardella, N. H., and M. H. Fugassa. 2009. Parasites in rodent coprolites from the historical archaeological site of Alero Mazquiarán, Chubut Province, Argentina. *Memoirs of the Institute Oswaldo Cruz* 104 (1): 37–42.

Saunders, G. R., M. N. Gentle, and C. R. Dickman. 2010. The impacts and management of foxes *Vulpes vulpes* in Australia. *Mammal Review* 40 (3): 181–211.

Schwarz, E., and H. K. Schwarz. 1943. The wild and commensal stocks of the house mouse *Mus musculus* Linnaeus. *Journal of Mammalogy* 24: 59–72.

Searle, J. B., C. S. Jones, İ Gündüz, M. Scascitelli, E. P. Jones, J. S. Herman, R. V. Rambau, L. R. Noble, R. J. Berry, M. D. Giménez, and F. Jóhannesdóttir. 2009. Of mice and (Viking?) men: Phylogeography of British and Irish house mice. *Proceedings of the Royal Society of London B* 276: 201–7.

Seed, A., N. Emery, and N. Clayton. 2009. Intelligence in corvids and apes: A case of convergent evolution? *Ethology* 115 (5): 401–20.

Serjeantson, D., and J. Morris. 2011. Ravens and crows in Iron Age and Roman Britain. *Oxford Journal of Archaeology* 30 (1): 85–107.

Serpell, J. A. 1988. The domestication and history of the cat. In *The Domestic Cat: The Biology of its Behaviour*, ed. D. C. Turner and P. Bateson, 151–58. Cambridge: Cambridge University Press.

Serpell, J. A. 1996. *In the Company of Animals: A Study of Human-Animal Relationships*. Cambridge: Cambridge University Press.

Shackleton, E. H. [1919] 1983. *South*. London: Century Hutchinson.

Shaw, L., D. Chamberlain, and M. Evans. 2008. The house sparrow *Passer domesticus* in urban areas: Reviewing a possible link between post-decline distribution and human socioeconomic status. *Journal of Ornithology* 149: 293–99.

Sheail, J. 1999. The grey squirrel (*Sciurus carolinensis*): A UK historical perspective on a vertebrate pest species. *Journal of Environmental Management* 55: 145–56.

Shipman, P. 2011. *The Animal Connection: A New Perspective on What Makes Us Human*. New York: W.W. Norton.

Shochat, E., P. S. Warren, and S. H. Faith. 2006. Future directions in urban ecology. *Trends in Ecology and Evolution* 21 (12): 661–62.

Silver, J. 1927. The introduction and spread of house rats in the United States. *Journal of Mammalogy* 8 (1): 58–60.

Sim, Y. H. 2005. *How to swing a mouse: Intersections of female and feline in Medieval Europe*. Oberlin University, Medieval Women Writers. http://www.oberlin.edu/library/friends/research awards/SIM.pdf (consulted 28 October 2011).

Skórka, P., R. Martyka, J. D. Wójcik, T. Babiary, and J. Skórka. 2006. Habitat and nest site selection in the common gull *Larus canus* in southern Poland: Significance of man-made habitats for conservation of an endangered species. *Acta Ornithologica* 4 (2): 137–44.

Slater, M. R., and S. Shain. 2005. Feral cats: An overview. In *The State of the Animals III*, ed. D. J. Salem, 43–53. Washington, DC: Humane Society Press.

Smith, B. D. 1998. *The Emergence of Agriculture*. New York: Scientific American Books.

Smith, D., and H. K. Kenward. 2011. Roman grain pests in Britain: Implications for grain supply and agricultural production. *Britannia* 42: 243–62.

Smith, R. C. 2001. The Emanuel Point ship: A 16th century vessel of Spanish colonization. *Trabalhos de Arqueologia* 18 (*Proceedings of the International Symposium of the Archaeology of Medieval–Modern Ships of Iberian-Atlantic Tradition*), 295–300. http://www.igespar.pt/media/uploads/trabalhosdearqueologia/18/23.pdf.

Smouse, P. E. 2000. Reticulation inside the species boundary. *Journal of Classification* 17 (2): 165–73.

Smout, T. C. 2003. The alien species in 20th century Britain: Constructing a new vermin. *Landscape Research* 28 (1): 11–20.

Sodikoff, G., ed. 2009. *The Anthropology of Extinction: Essays on Culture and Species Death*. Bloomington: Indiana University Press.

Sol, D., D. M. Santos, and M. Cuadrado. 2000. Age-related feeding site selection in urban pigeons (*Columba livia*): Experimental evidence of the competition hypothesis. *Canadian Journal of Zoology* 78 (1): 144–49.

Sol, D., S. Timmermans, and L. Lefebvre. 2002. Behavioural flexibility and invasion success in birds. *Animal Behaviour* 63: 495–502.

Soldatini, C., Y. V. Albores-Barajas, D. Mainardi, and P. Monaghan. 2008. Roof nesting by gulls for better or worse? *Italian Journal of Zoology* 75 (3): 295–303.

Starkovich, B. M., and M. Stiner. 2009. Hallan Çemi Tepesi: High-ranked game exploitation alongside intensive seed processing at the Epipalaeolithic-Neolithic transition in southeastern Turkey. *Anthropozoologica* 44 (1): 41–61.

Steffen, W., P. J. Crutzen, and J. R. McNeill. 2007. The Anthropocene: Are humans now overwhelming the great forces of nature? *AMBIO: Journal of the Human Environment* 36: 614–21.

Stewart, J. R. 2004. Neanderthal-modern human competition? A comparison between the mammals associated with Middle and Upper Palaeolithic industries in Europe during OIS3. *International Journal of Osteoarchaeology* 14: 178–89.

Strubbe, D., E. Matthysen, and C. H. Graham. 2010. Assessing the potential impact of invasive ring-necked parakeets *Psittacula krameri* on native nuthatches *Sitta europaea*. *Journal of Applied Ecology* 47 (3): 549–57.

Sukopp, H. 2008. On the early history of urban ecology in Europe. In *Urban Ecology*, ed. J. Marzluff et al., 79–98. New York: Springer.

Summa, M., C.-H. von Bonsdorff, and L. Maunula. 2012. Pet dogs—A transmission route for human noroviruses? *Journal of Clinical Virology* 53 (3): 244–47.

Suzuki, H., T. Shimada, M. Terashima, K. Tsuchiya, and K. Aplin. 2004. Temporal, spatial, and ecological modes of evolution of Eurasian *Mus* based on mitochondrial and nuclear gene sequences. *Molecular Phylogenetics and Evolution* 33: 626–46.

Svensson, J. R., M. Lindegarth, M. Siccha, M. Lenz, M. Molis, M. Wahl, and H. Pavia. 2007. Maximum species richness at intermediate frequencies of disturbance: Consistency among levels of production. *Ecology* 88 (4): 830–38.

Sykes, N. M. 2004. The introduction of fallow deer to Britain: A zooarchaeological perspective. *Environmental Archaeology* 9 (1): 75–83.

Sykes, N. M. 2010. European fallow deer. In *Extinctions and Invasions: A Social History of British Fauna*, ed. T. P. O'Connor, and N. M. Sykes, 51–58. Oxford: Windgather Press.

Tabor, R. K. 1983. *The Wild Life of the Domestic Cat*. London: Arrow Books.

Tarsitano, E., G. Greco, N. Decaro, F. Nicassio, M. S. Lucente, C. Buonavoglia, and M. Tempestra. 2010. Environmental monitoring and analysis of faecal contamination in an urban setting in the city of Bari (Apulia Region, Italy): Health and hygiene implications. *International Journal of Environmental Research and Public Health* 7: 3972–86.

Taylor, K. 2009. Biological flora of the British Isles: *Urtica dioica* L. *Journal of Ecology* 97 (6): 1436–58.

Tchernov, E. 1984. Commensal animals and human sedentism in the Middle East. In *Animals and Archaeology: 3. Early Herders and Their Flocks*, ed. J. Clutton-Brock and G. Grigson, 91–115. Oxford: British Archaeological Reports, International Series 202.

Tchernov, E. 1991a. Biological evidence for human sedentism in Southwest Asia during the Natufian. In *The Natufian Culture in the Levant*, ed. O. Bar-Yosef and F. R. Valla, 315–40. Archaeological Series 1. Ann Arbor: International Monographs in Prehistory.

Tchernov, E. 1991b. Of mice and men: Biological markers for long-term sedentism: A reply. *Paléorient* 17 (1): 153–60.

Tchernov, E. 1993. The effect of sedentism on the exploitation of the environment in Southern Levant. In *Exploitation des animaux sauvages à travers le temps*, ed. J. Desse and F. Audoin-Rouzeau, 137–59. Juan-les-Pins, France: Editions APDCA.

Tchernov, E. 1994. *An Early Neolithic Village in the Jordan Valley*. Part 2: *The Fauna of Netiv Hagdud.* Cambridge, MA: Peabody Museum of Archaeology and Ethnology, Harvard University.

Tchernov, E., and F. Valla. 1997. Two new dogs and other Natufian dogs from the Southern Levant. *Journal of Archaeological Science* 24: 65–95.

Thompson, J. N. 1989. Concepts of coevolution. *Trends in Ecology and Evolution* 4 (6): 179–83.

Thornton, R. K. R. 1997. *John Clare: Everyman's Poetry*. London: J.M. Dent.

Tichy, H., Z. Zaleska-Rutczynska, C. O'Huigin, F. Figueroa, and J. Klein. 1994. Origins of the North American house mouse. *Folia biologica* 40 (6): 483–96.

Todd, N. B. 1978. An ecological, behavioural genetic model for the domestication of the cat. *Carnivore* 1: 52–60.

Tomek, T., and Z. Bocheński. 2006. Weichselian and Holocene bird remains from Komarowa Cave, Central Poland. *Acta zoological cracowiensia* 48A: 43–65.

Toškan, B., and B. Kryštufek. 2006. Noteworthy rodent records from the Upper Pleistocene and Holocene of Slovenia. *Mammalia* 70: 98–105.

Traweger, D., R. Travnitzky, C. Moser, C. Walzer, and G. Bernatzky. 2006. Habitat preferences and distribution of the brown rat (*Rattus norvegicus* Berk.) in the city of Salzburg (Austria): Implications for an urban rat management. *Journal of Pest Science* 79: 113–25.

Trut, L., I. Oskina, and A. Kharlamova. 2009. Animal evolution during domestication: The domesticated fox as a model. *Bioessays* 31 (3): 349–60.

Tucker, P. K., B. K. Lee, B. L. Lundgren, and E. M. Eicher. 1992. Geographic origin of the Y chromosome in "old" inbred strains of mice. *Mammalian Genome* 3 (5): 254–61.

Twiss, K. C. 2007. The zooarchaeology of Tel Tif'Dan (Wadi Fidan 001), Southern Jordan. *Paléorient* 33 (2): 127–45.

Usher, M. B., T. J. Crawford, and J. L. Banwell. 1992. An American invasion of Great Britain: The case of the native and alien squirrel (*Sciurus*) species. *Conservation Biology* 6 (1): 108–15.

Van der Mieroop, M. 1999. *The Ancient Mesopotamian City.* Oxford: Oxford University Press.

Van Heezik, Y., A. Smyth, A. Adams, and J. Gordon. 2010. Do domestic cats impose an unsustainable harvest on urban bird populations? *Biological Conservation* 143 (10): 121–30.

Van Schaik, C. P., M. Ancrenaz, G. Borgen, B. Galdikas, C. D. Knott, I. Singleton, A. Suzuki, S. Utami, and M. Merrill. 2003. Orangutan cultures and the evolution of material culture. *Science* 299: 102–5.

Van Schalkwyk, J. A., C. J. Meyer, A. Pelser, and I. Plug. 1995. Images of the social life and household activities at Melrose House. *Research by the National Cultural Museum* 4: 81–99.

Vázquez-Domínguez, E., G. Ceballos, and J. Cruzado. 2004. Extirpation of an insular subspecies by a single introduced cat: The case of the endemic deer mouse *Peromyscus guardia* on Estanque Island, Mexico. *Oryx* 38: 347–50.

Veitch, C. R. 2002. Eradication of Norway rats (*Rattus norvegicus*) and house mouse (*Mus musculus*) from Motuihe Island, New Zealand. In *Turning the Tide: The Eradication of Invasive Species*, ed. C. R. Veitch and M. N. Clout, 353–56. Cambridge: IUCN SSC Invasive Species Specialist Group.

Vellanoweth, R. L., B. G. Bartelle, A. F. Ainis, A. C. Cannon, and S. J. Schwartz. 2008. A double dog burial from San Nicolas Island, California, USA: Osteology, context, and significance. *Journal of Archaeological Science* 35 (12): 3111–23.

Vigne, J-D., J. Guilane, K. Debue, L. Haye, and P. Gérard. 2004. Early taming of the cat in Cyprus. *Science* 304: 259.

Vijman, V., and K. A. I. Nekaris. 2010. Effects of deforestation and levels of tolerance towards commensal primates (Cercopithecidae) in Sri Lanka. *International Journal of Pest Management* 56 (2): 153–58.

Voigt, E. A., and A. von den Driesch. 1984. Preliminary report on the faunal assemblage from Ndondondwane, Natal. *Annals of the Natal Museum* 26: 95–104.

Vuorisalo, T., H. Anderson, T. Hug, R. Lahtinen, H. Laaksonen, and E. Leihikoinen. 2003. Urban development from an avian perspective: Causes of hooded crow (*Corvus corone cornix*) urbanisation in two Finnish cities. *Landscape and Urban Planning* 62 (2): 69–87.

Waite, T. A., A. K. Chhangani, L. G. Campbell, L. S. Rajpurohit, and S. M. Mohnot. 2007. Sanctuary in the city: Urban monkeys buffered against catastrophic die-off during ENSO-related drought. *EcoHealth* 4: 278–86.

Wallace, A. R. 1898. *The Wonderful Century: Its Successes and Its Failures.* London: Swan Sonnenschein & Co.

Wang, W-M., J-L. Ding, J-W. Shu, and W. Chen. 2010. Exploration of early rice farming in China. *Quaternary International* 227 (1): 22–28. doi: 10.1016/j.quaint.2010.06.007.

Warren, Y. 2008. Crop-raiding baboons (*Papio anubis*) and defensive farmers: A West African perspective. *West African Journal of Applied Ecology* 14: 1–11.

Watkins, T. 2010. New light on Neolithic revolution in south-west Asia. *Antiquity* 84: 621–34.

Watson, G. E. 1975. *Birds of the Antarctic and Sub-Antarctic.* Washington, DC: American Geophysical Union.

Watson, N. L., and M. Weinstein. 1993. Pet ownership in relation to depression, anxiety, and anger in working women. *Anthrozoos* 6 (2): 135–38.

Waudby, H. 2010. Population characteristics of house mice (*Mus musculus*) on southern Yorke Peninsula, South Australia. *Australian Mammalogy* 31 (2): 111–15.

Webster, J. C. 2007. Missing cats, stray coyotes: One citizen's perspective. *Wildlife Damage Management Conferences—Proceedings*, 74–116. Lincoln: University of Nebraska. http://digitalcommons.unl.edu/icwdm_wdmconfproc/78.

Weisdorf, J. 2005. From foraging to farming: Explaining the Neolithic Revolution. *Journal of Economic Surveys* 19 (4): 561–86.

Weissbrod, L., T. Dayan, D. Kaufman, and M. Weinstein-Evron. 2005. Micromammal taphonomy of el-Wad Terrace, Mount Carmel, Israel: Distinguishing cultural from natural depositional agents in the Late Natufian. *Journal of Archaeological Science* 2 (1): 1–17.

Welford, M. R., and B. H. Bossack. 2009. Validation of inverse seasonal peak mortality in medieval plagues, including the Black Death, in comparison to modern *Yersinia pestis*–variant diseases. *PLoS ONE* 4 (12): e8401.

Welford, M. R., and B. H. Bossack. 2010. Revisiting the Medieval Black Death of 1347–1351: Spatiotemporal dynamics suggestive of an alternate causation. *Geography Compass* 4 (6): 561–75.

White, C. 2005. Hunters ring dinner bell for ravens: Experimental evidence of a unique foraging strategy. *Ecology* 86 (4): 1057–60.

White, G., and F. Greenoak. 1988. *The Journals of Gilbert White.* Vol. 2. London: Century.

White, T., B. Asfaw, Y. Beyenne, Y. Haile-Selassie, C. O. Lovejoy, G. Suwa, and G. WoldeGabriel. 2009. *Ardipithecus ramidus* and the paleobiology of early hominids. *Science* 326 (5949): 75–86.

Willcox, G., R. Buxo, and L. Herveux. 2009. Late Pleistocene and early Holocene climate and the beginnings of cultivation in northern Syria. *The Holocene* 19 (1): 151–58.

Wilson, D. E., and R. A. Mittermeier, eds. 2009. *Handbook of the Mammals of the World.* Vol. 1, *Carnivores.* Barcelona: Lynx Edicions.

Wilson, E. O. 1984. *Biophilia: The Human Bond with Other Species.* Cambridge, MA: Harvard University Press.

Wilson, P. J. 2007. Agriculture or architecture? The beginnings of domestication. In *Where the Wild Things Are Now*, ed. R. Cassidy and M. Mullen, 101–21. Oxford: Berg.

Woodward, P. J., S. M. Davies, and A. H. Graham. 1993. *Excavations at the Old Methodist Chapel and Greyhound Yard, Dorchester, 1981–4*. Dorset Natural History and Archaeology Society Monograph 12. Dorchester, UK: Dorset Natural History and Archaeology Society.

Wright, J. P., and C. G. Jones. 2006. The concept of organisms as ecosystem engineers ten years on: Progress, limitations, and challenges. *Bioscience* 56: 203–9.

Wycoll, G., and D. Tangri. 1991. The origins of commensalism and human sedentism. *Paléorient* 17 (2): 157–59.

Yamagiwa, J., and D. A. Hill. 1983. Intraspecific variation in the social organization of Japanese macaques: Past and present scope of field studies in natural habitats. *Primates* 39 (3): 257–73.

Yeshurun, R., G. Bar-Oz, and M. Weinstein-Evron. 2009. The role of foxes in the Natufian economy: A view from Mount Carmel, Israel. *Before Farming* 1 (3): 1–15.

Yu, H.-T., and Y.-H. Peng. 2002. Population differentiation and gene flow revealed by microsatellite DNA markers in the House Mouse (*Mus musculus castaneus*) in Taiwan. *Zoological Science* 19: 475–83.

Zeder, M. A. 2006. Central questions in the domestication of plants and animals. *Evolutionary Anthropology* 15: 105–17.

Zeuner, F. E. 1963. *A History of Domesticated Animals*. London: Hutchinson.

Zhang, P., K. Watanabe, and T. Elishi. 2007. Habitual hot-spring bathing by a group of Japanese macaques (*Macaca fuscata*) in their natural habitat. *American Journal of Primatology* 69 (12): 1425–30.

Index

humans + animals
in the Napa Valley